RAND NATIONAL DEFENSE RESEARCH INSTITUTE

T0289424

An Enhanced Capability to Model How Compensation Policy Affects U.S. Department of Defense Civil Service Retention and Cost

David Knapp, Beth J. Asch, Michael G. Mattock, James Hosek

Prepared for the Office of the Secretary of Defense,
Personnel and Readiness/DCPAS/HROPS (Compensation)

For more information on this publication, visit www.rand.org/t/RR1503

Library of Congress Cataloging-in-Publication Data
ISBN: 978-0-8330-9639-5

Published by the RAND Corporation, Santa Monica, Calif.
© Copyright 2016 RAND Corporation
RAND® is a registered trademark.

Cover design by Tanya Maiboroda
Cover images: Man in foreground, ©ra2 studio; Question marks, ©Jezper; Dollar sign, ©Fotolia365;
Business people hierarchy, ©ag visual; Arrows, ©gearstd.

Support RAND
Make a tax-deductible charitable contribution at
www.rand.org/giving/contribute

www.rand.org

Preface

Retention is a critical part of workforce management, as is the strategic use of compensation to influence the size and composition of that workforce. To manage the U.S. Department of Defense (DoD) civilian workforce and use compensation effectively and efficiently, planners and policymakers must have the capability to rigorously assess the retention effects and cost implications of changing the level and structure of compensation. That capability needs to have a solid foundation in the literature and theories of how personnel make retention decisions over their careers, be firmly grounded empirically with data on retention decisions at the individual level of personnel careers, and include a capacity to perform simulations of the retention and cost effects of relevant policy changes. The dynamic retention model provides such a capability.

Earlier RAND work (Asch, Mattock, and Hosek, 2014a) developed a prototype dynamic retention model for DoD civilian employees, and the current research greatly extends the capability of that model by obtaining new parameter estimates for a broader range of retention data, developing new estimates of federal civilian and private-sector pay profiles, adding a costing capability, and testing for possible change in the preference for DoD civilian employment among entry cohorts of workers hired in the 1990s. The report also demonstrates the expanded modeling capability by considering the effects on retention and cost of the higher employee retirement contributions mandated under Pub. L. 112-96, Middle Class Tax Relief and Job Creation Act of 2012, Section 5001. The model development and policy analysis should be of interest to the policy community concerned with the effectiveness of federal compensation, as well as the research community analyzing human resource and personnel issues.

This research was sponsored by the Office of the Under Secretary of Defense for Personnel and Readiness, Civilian Personnel Policy, Defense Civilian Personnel Advisory Service, conducted within the Forces and Resources Policy Center of the RAND National Defense Research Institute, a federally funded research and development center sponsored by the Office of the Secretary of Defense, the Joint Staff, the Unified Combatant Commands, the Navy, the Marine Corps, the defense agencies, and the defense Intelligence Community.

For more information on the RAND Forces and Resources Policy Center, see www.rand.org/nsrd/ndri/centers/frp or contact the director (contact information is provided on the web page).

Contents

Figures and Tables

Figures

Tables

Summary

Given the size of the federal workforce and, in the context of the U.S. Department of Defense (DoD), its contribution to military readiness, planners and policymakers must have the capability to understand how changes in compensation and personnel policy affect that workforce. The dynamic retention model (DRM) can provide such a capability. Our earlier analysis, summarized in Asch, Mattock, and Hosek, 2014a, represented a step toward developing that capability. The research summarized in this report further develops and refines this capability. Specifically, we estimate the model for entry cohorts entering between 1988 and 2000 (rather than just 1988, like in the previous work), allow the taste distributions to differ between veterans and nonveterans, refine the pay profiles that are key inputs to the model, and extend the DRM simulation capability to compute the changes in personnel costs associated with changes in compensation policy. The improved modeling capability will enable DoD to assess the effects that compensation changes have on workforce retention and costs.

Given the extension of our model to additional entry cohorts, we began our analysis by considering how the demographic composition of these entry cohorts changed between 1988 and 2000. Characteristics do indeed differ across the entry cohorts. One of the most dramatic changes is the increase in the percentage of entrants who are veterans. There is also an associated increase in the mean age of entrants given that veterans are older and a majority of veterans who enter civil service are military retirees. The percentage of entrants who are veterans increased from about 5 percent in 1991 to about 24 percent in 1994. We find a shift toward higher entry grades, controlling for educational attainment, across the 1988 to 2000 entry cohorts. This shift persists even among nonveteran entrants. We also find a shift toward a more-educated DoD civilian workforce and a rising percentage of employees who are female and a rising percentage of employees who are minorities for both veteran and nonveteran entrants. The changing demographics might be a result of the drawdown policies of the 1990s that targeted younger personnel in lower grades (Chu and White, 2000) or technological change that is biased toward positions with higher starting pay grades and more education.

As part of the DRM, people choose whether to stay in the civil service partly based on how pay in the civil service compares with external expected pay. Given the longitudinal nature of the DRM, we conducted an analysis of whether pay profiles estimated from American Community Survey (ACS) cross-sectional data accurately represented longitudinal pay profiles by birth cohort. Use of the ACS data is preferred given its greater sample size; however, the ACS did not start until 2000. To estimate longitudinal pay profiles by birth cohort, we generated synthetic birth cohorts using Current Population Survey (CPS) data for 1966 to the present.[1] Our findings indicate that the pay profiles estimated for private workers were highly similar for ACS data and CPS synthetic birth cohorts. The results were also highly similar for federal civil service workers for birth cohorts born in 1963 or later, but the CPS profiles were higher than ACS profiles for the 1943 birth cohort after age 40 and somewhat higher for the 1953 birth cohort after that age. We view these results as somewhat anomalous. Overall, we conclude that pay profiles for both private workers and federal civil servants from the ACS cross-sectional data are valid to use in estimating the DRM.

Several conclusions emerged from our estimation of the DRM by entry cohort and our estimation of the DRM combining entry cohorts. First, we continue to find excellent model fits in our extended analysis. The good fit between our model predictions and the actual data gives us confidence about the model estimates.

Second, we find that the taste distributions for civil service differ between veterans and nonveterans. We estimate higher mean taste among veterans but more-homogeneous tastes (lower standard deviation), and these estimated differences are statistically significant from 0. The differences between veterans and nonveterans are unsurprising, especially given that many veterans who enter civil service are military retirees. Military retirees have already spent at least 20 years working in support of the military mission and have revealed themselves to have a high taste for DoD employment. As a group that has self-selected already on their taste for military service, their tastes for DoD civil employment are also relatively homogeneous. This conclusion is consistent with DRM estimates for military personnel, in which we find that mean taste increases with years of service as lower-taste personnel separate and higher-taste personnel stay, and the standard deviation of taste decreases because personnel who stay have more-homogeneous tastes (Mattock, Asch, et al., 2014).

Third, we find that the estimated positive mean taste for civil service employment found for the 1988 entry cohort in Asch, Mattock, and Hosek, 2014a, was anomalous. We again find a positive mean taste for the 1988 entry cohort, but we also find that, among entry cohorts entering after 1989, estimated mean tastes are generally either negative or near 0. In the model, where we combine entry cohorts entering between 1992 and 2000, the estimated mean taste is –$4,160 among nonveterans and –$1,210 among veterans. Other parameters also change somewhat. Consequently, responsiveness to compensation policy changes also differs somewhat from that in the earlier study. For example, for the 1988 entry cohort, a 1-percent drop in real salaries results in a 2.1-percent drop in the civil service workforce size that is retained in the steady state. But, using the estimated DRM that combines the 1992–2000 entry cohorts, we find that a 1-percent drop in real pay would reduce workforce size by 3.7 percent in the steady state. Thus, the extended model shows a larger retention response than our earlier model would have predicted.

Finally, we explored whether the taste distribution has changed for more-recent entry cohorts. Some have argued that the career aspirations and cultural attitudes of millennials, generation X, baby boomers, and Depression-era and wartime generations (also known as the Silent Generation) differ. The common argument is that younger generations evaluate their labor market opportunities differently from older generations and have different priorities for their careers. For example, a report on millennials by PricewaterhouseCoopers, 2011, showed that, when millennials were asked, "Which of the following factors most influenced your deci-

---

[1] In lieu of panel data, a synthetic cohort appends nationally representative cross-section samples to create panel data that reflect national averages. Deaton, 1985, provides further detail on the use of synthetic panels.

sion to accept your current job?" the top categories were opportunity for personal development (65 percent) and the reputation of the organization (36 percent). Only 21 percent said that salary was a significant influencer in taking the current job. According to the Pew Research Center, 68 percent of millennials were in the civilian labor force between the ages of 18 and 33, compared with 78 percent for the previous three generations for the same age range (Pew Research Center, 2015). Alternatively, it could be that calendar year, not birth year, is more relevant for decisionmaking. For example, one could argue that changes in societal attitudes and culture over time affect all people in a similar fashion, regardless of birth year. Thus, the tastes of all people entering public service in 2000 might differ from tastes of those entering in 1988, regardless of age.

To explore whether we find evidence of changing tastes across entry cohorts, we estimate a combined DRM using data for all entry cohorts that includes dummy variables allowing the mean taste parameter to differ for each entry calendar year. The dummy variables allow us to test whether mean taste shifted across entry calendar year. We estimate this combined model for a specific demographic subset of the data to limit the possibility that the observed differences are due to changes in the demographic composition of the entry cohorts over time. In particular, we estimate a combined model for male nonveterans, ages 30 or younger, with only bachelor's degrees, entering between 1992 and 2000.

We find evidence that mean taste has changed across entry cohorts because some differences across entry cohorts in mean tastes are statistically significant. We find that mean taste for civil service, relative to the 1992 entry cohort, declines significantly for the 1993–1995 entry cohorts, returns to the 1992 level for entry cohorts entering in 1996 to 1998, and increases significantly above this level for entry cohorts entering in 1999 and 2000. Again, these results are for the subset of employees defined to have homogeneous observed demographic characteristics. It is unclear how important these differences in mean tastes are. We find that retention responsiveness to a 1-percent real decrease in federal salaries varies from –5.3 percent to –3.2 percent. So, the differences across entry cohorts seem to translate into some difference in pay responsiveness.

For the purposes of policy analysis and conducting simulations of the retention and cost effects of changing compensation, we needed to select a set of estimates, but we recognize that the selection could affect our results, given the differing estimates across entry cohorts. To address this concern, we selected the DRM estimates that combine entry cohorts—specifically, the 1992–2000 entry cohorts. This model fits the data very well and accounts for differences in the taste distribution between veterans and nonveterans.

To demonstrate the simulation capability of our extended model, especially the costing capability that we developed, we analyzed the impact that the higher employee contribution rates mandated under Pub. L. 112-96, Middle Class Tax Relief and Job Creation Act of 2012, Section 5001, have on DoD civilian retention. These contributions help cover the cost of the Federal Employees Retirement System (FERS) defined benefit plan or basic plan. Our analysis of the policy change's cost impact incorporates the retention response insofar as employees change their retention behavior when mandated contributions increase. We consider different cases regarding how employees might change their consumption and saving behavior when mandated contributions increase, allowing us to consider a range of retention and cost impacts.

We find that, depending on how employees alter their consumption and saving behavior, the change in workforce size retained could vary from 0 to –12.2 percent. The former figure corresponds to the case in which employees respond to the mandate by reducing their savings (other than Thrift Savings Plan [TSP] retirement contributions), while the latter corresponds to the case in which they respond by reducing current consumption (i.e., by reducing the amount of money from each paycheck available to be spent). The decrease in DoD's total personnel costs per employee (defined as the salary and retirement costs to DoD of employing its General Schedule workforce with at least a bachelor's degree and covered by FERS, on a per-employee basis) varies between 3 and 5 percent. The 3-percent decrease corresponds to the case in which employees respond to the mandate by reducing their savings (other than TSP retirement contributions). The 5-percent decrease corresponds to when employees reduce their TSP contributions to cover the higher mandated contributions.

Our simulation results confirm the intuition behind the policy change. Mandated employee contributions were increased to reduce the cost to DoD (and other federal agencies) of providing the FERS basic plan. However, our analysis incorporates the retention response to the policy change and shows that additional cost savings might occur because the policy changes the experience mix of the force that is retained, thereby affecting salary and retirement costs per employee. Furthermore, insofar as employees respond to the policy by reducing their TSP contributions, our analysis also incorporates the additional cost savings to DoD of lower TSP matching contributions. Thus, the simulation capability provides a more nuanced assessment of cost changes as a result of compensation policy changes. We emphasize that these are cost savings per se and do not consider how changes in the size or experience mix of the workforce affect its productive capability; a less costly, more junior workforce might be less productive, for instance. It also does not incorporate how the policy change affects hiring. These are potential areas for future research.

Acknowledgments

We gratefully acknowledge the support and insightful comments of Christopher P. Lynch, chief, Strategic Compensation Branch and International Pay Policy, Defense Civilian Personnel Advisory Service Human Resource Operational Programs and Advisory Services Compensation. We also thank Sharilyn Spicher of the Defense Human Resources Activity for her helpful input. We appreciate the help of the Defense Manpower Data Center in providing data. We would like to thank Whitney Dudley of RAND for programming assistance. We also greatly appreciate the helpful comments and input we received from the reviewers, Matthew S. Goldberg, deputy assistant director of the National Security Division at the Congressional Budget Office, and Matthew D. Baird of RAND.

Abbreviations

ACS	American Community Survey
CI	confidence interval
CPS	Current Population Survey
DoD	U.S. Department of Defense
DRM	dynamic retention model
FERS	Federal Employees Retirement System
GS	General Schedule
OPM	U.S. Office of Personnel Management
SD	standard deviation
SE	standard error
TSP	Thrift Savings Plan

CHAPTER ONE

Introduction

Retention is a critical part of workforce management, as is the strategic use of compensation to influence the size and composition of that workforce. To manage the workforce and use compensation effectively and efficiently, planners and policymakers must have the capability to rigorously assess the retention effects and cost implications of changing the level and structure of compensation. That capability needs to have a solid foundation in the literature and theories of how personnel make retention decisions over their careers, be firmly grounded empirically with data on retention decisions at the individual level of personnel careers, and include a capacity to perform simulations of the retention and cost effects of relevant policy changes. The dynamic retention model (DRM) provides such a capability.

The DRM is a structural, stochastic, dynamic discrete-choice model of individual behavior in which people make retention decisions under uncertainty during their careers and have unique or heterogeneous tastes. In the model, employees make retention decisions throughout their careers about whether to remain in or leave the organization. The DRM incorporates a taste factor that captures a person's preference for working for the organization relative to the external market and includes persistent nonmonetary and monetary factors not otherwise included in the model. People are forward-looking. They have expectations about the likelihood of future events and know their eligibility for future benefits, such as retirement benefits, conditional on the outcomes of these events. Their current decisions incorporate both these expectations and their past employment histories.

The DRM approach has been used in a variety of settings, such as studying the retention of Air Force pilots (Gotz and McCall, 1984; Mattock and Arkes, 2007), reserve retirement reform (Asch, Hosek, and Mattock, 2013), and military retirement reform (Asch, Hosek, and Mattock, 2014; Asch, Mattock, and Hosek, 2015).

In Asch, Mattock, and Hosek, 2014a, and Asch, Mattock, and Hosek, 2014b, we extended the DRM to federal civil service employment, using 24 years of data on defense civilians, and simulated the effects that pay freezes and unpaid furloughs have on civil service retention. That study focused on only one entry cohort of entering civilians (1988), did not consider demographic or other relevant differences across employees in that entry cohort, and considered only retention and not costs. In this report, we focus on the new data and extensions to the DRM needed to analyze the retention and cost effects of recent proposals to alter the Federal Employees Retirement System (FERS).

In particular, we increase the model's functionality and provide an assessment of the retention and cost effects of recent proposals to change FERS. We extend the model in four ways. First, we consider entry cohorts for 1988 to 2000. Additional entry cohorts allow us to understand how the civil service has changed over time and examine and incorporate into the estimates important sources of demographic and other differences across these entry cohorts. We estimate separate models for each entry cohort, as well as a combined model. This highly flexible approach reveals the extent to which the models for individual entry cohorts produce different results and explore possible reasons for those differences. A comparison of the combined model with the separate models serves as a probe of the robustness of the combined model. Second, within entry cohorts, we consider group differences in tastes for military service, focusing specifically on differences between veteran and nonveteran employees. As we discuss in Chapter Two, the percentage of entrants who are under the federal civil service General Schedule (GS) and are veterans increased dramatically between 1991 and 1994 and remained high thereafter.[1] Third, we extend to the civil service a method initially developed for analyzing cost changes resulting from changes in military compensation policy. The method, adapted to the civil service, allows the calculation of (1) changes in current compensation costs, e.g., employee salaries, and (2) changes in the costs associated with deferred compensation, e.g., employer contributions to the Thrift Savings Plan (TSP) and the normal costs of the FERS defined benefit plan. Finally, we refine the estimates of how pay varies over a federal career versus a career in private employment. The estimated pay profiles are key inputs to the DRM. The new pay estimates account for top-coding of income data in the American Community Survey (ACS), which could potentially bias the pay profile estimates. As we discuss later, top-coding is the use of specific thresholds in publicly released data to protect the privacy of high-income people. If not taken into account, top-coding can lead to biased estimates of the pay profiles. In addition, the pay analysis addresses an important potential shortcoming of the ACS data—namely, the ACS pay regressions essentially capture the age–earning relationship in the cross-section, whereas the DRM ideally needs expected earnings over an entry cohort's career. By conducting a parallel analysis with many years of Current Population Survey (CPS) data that we use to estimate each entry cohort's earnings over time, we confirm that the ACS results accurately represent entry cohort earning profiles.[2] That is, the pay structure of both federal and private workers has been stable over time, leading to the result that the age–earning relationship in the cross-section is quite similar to the age–earning relationship of the entry cohorts studied. Like in our earlier work, we apply the estimated DRM to simulate the effect of a policy change mandating workers to increase their TSP contributions. This analysis demonstrates the capability of the extended DRM for civil service to quantify the retention effects of compensation policy changes and compute the policy costs in terms of current and deferred compensation.

The report is organized as follows. Chapter Two provides an overview of our data, including tabulations describing the demographic characteristics of entry cohorts between 1988 and 2000, a discussion of how they have evolved during this time frame, and implications for civil

---

[1]   The GS pay scale covers approximately 1.5 million of the 2.7 million nonmilitary federal employees.

[2]   The ACS has more observations than the CPS, but the ACS has a shorter time series. The CPS, covering 1962 to 2014, allows us to create synthetic cohort earning profiles. As we describe in Chapter Four, we compare those CPS cohort earning profiles to validate that they reflect CPS cross-sectional earning profiles. We then compare the CPS cross-sectional earning profiles and the ACS cross-sectional earning profiles and confirm that there are no systematic differences in the cross-sectional earning profiles across surveys. We opt to use the ACS cross-sectional earning profiles because of the larger sample size, but the analysis of cohort differences in the CPS data allows us to assess the presence of intergenerational differences in earnings. Systematic differences in a cohort's private versus federal earning profiles could create differences in the value of deferred and current compensation that could alter people's life-cycle choices of whether or not to stay in the federal civil service.

service retention. In Chapter Three, we provide a brief overview of the DRM—summarizing our earlier model and describe our extensions to the model in light of the demographic and entry cohort differences described in Chapter Two. Chapter Three also describes our methodology for calculating the changes in cost associated with changes in compensation policy. Chapter Four discusses our refinement of the pay profiles that we include in the DRM to account for civil service and private salary opportunities. The refinement accounts for top-coding in the data used to estimate the pay profiles. In that chapter, we also explore how pay profiles vary across birth cohorts and the value of using a birth cohort–specific versus a single pay profile in our modeling. Chapter Five shows our model estimates by cohort, accounting for taste differences between veteran and nonveterans, our estimates for models that combine cohorts, and our exploration of how and why model estimates vary across cohorts. Using what we consider the best working estimates from Chapter Five, Chapter Six presents simulation results. Our simulations consider several recent changes or proposed changes to the defined benefit plan under FERS. Specifically, we consider the retention and cost effects of increases in mandated employee contributions to TSP. Chapter Seven summarizes our conclusions and discusses areas for future research. We also include two appendixes: Appendix A provides the details for estimating the pay profiles, and Appendix B reestimates the models from Chapter Five with a fixed value of the discount factor.

CHAPTER TWO

Entry Cohort Characteristics

In this chapter, we describe the characteristics of each entry cohort of GS U.S. Department of Defense (DoD) civilians with bachelor's degrees or more of education and how those characteristics varied between 1988 and 2000. This time frame is important because it reflects a period of substantial technological change in which personal computers were becoming more important and greater emphasis was being placed on higher education in the DoD civil service. As we document later in the chapter, the percentage of these new civil service entrants with at least master's degree increased from 17 percent in 1988 to 31 percent in 1994, although it fell to 25 percent in 2000. The tabulations shown in this chapter allow us to better understand how the college-educated civil service has changed over time and provide information on possible important demographic differences across these entry cohorts that should be incorporated into the DRM. We explore the implications for the DRM in the next chapter. We present tabulations of the characteristics for each entry cohort and how they have varied across entry cohort entry years. We begin by first describing the data we use and the analysis file we developed on DoD civilian entrants and their subsequent retention. We use this file to estimate the DRM and tabulate entry cohort characteristics.

Data on Civil Service Entry Cohorts

To estimate the DRM, we built longitudinal data files that track the individual careers of employees who entered the DoD civil service between 1988 and 2000. We constructed these files from the Defense Manpower Data Center master file and transaction file data on DoD civil service personnel. The longitudinal file tracks these personnel from entry until 2012 or separation, whichever came first. Thus, we were able to track the careers of DoD civil service personnel over a 24-year period.

We excluded some observations from the analysis. From 1988 to 2000, there were 403,103 entrants into the DoD civil service. The project sponsor requested that we focus on GS employees with at least bachelor's degrees, so we excluded entrants who did not meet this criterion.[1] This resulted in a sample of 74,238 entrants for 1988 to 2000. We also excluded temporary workers and those who worked less than full time, were considered inactive or seasonal,

---

[1] Sample selection was based on these characteristics at entry to the DoD civil service. If someone changed pay schedules (as was common from 2006 to 2010 under the National Security Personnel System) or changed work schedule, they were retained in the sample as long as they continued to work for DoD. Additionally, we continued to leave out people who were selected out of the sample based on selection criteria in their first year of service (e.g., part-time workers) but satisfied the selection criteria in later years (e.g., part-time workers transitioning to full-time workers).

or were not covered by FERS. This reduced the sample to 70,075. GS employees with at least bachelor's degrees enter the civil service at a range of ages, as we show in the next section. We limit the entry ages to between 25 and 52, capturing 96.9 percent of full-time entrants on the GS with at least bachelor's degrees.

FERS, which was introduced in 1987, covers all the people in the data we used to estimate the DRM. It consists of three parts: a defined benefit plan that bases retirement benefits on the employee's earnings and years of service, Social Security coverage, and a defined contribution plan (the TSP). All new entrants with less than five years of service are automatically placed under FERS. Because the people in our analysis had no prior service in 1988, they fall into this category.

The longitudinal files we constructed for each entry cohort are similar to those constructed for the 1988 entry cohort in Asch, Mattock, and Hosek, 2014a. As discussed in that report, we recognize that people can gain additional education while they are in the civil service and can switch occupations. However, for reasons discussed in the earlier report, we opted to use the entry education level. Also, like in the earlier analysis, we excluded people who separated and later returned because this behavior is not currently built into the DRM. This further reduces the sample to 61,129 entrants between 1988 and 2000, capturing 87.2 percent of full-time entrants on the GS with at least bachelor's degrees. By excluding the possibility of returning to the civil service, we made the decision to leave more costly because the person's FERS pension is frozen at separation. This has the effect of inducing greater retention in the model, which is relevant when considering the results of our analysis in Chapter Six.

Entry Cohort Characteristics

Figure 2.1 shows the mean entry age by cohort entry date for veterans and nonveterans. Mean age is relatively stable for each of these groups, but veterans are a dozen years older than nonveterans, about 42 versus about 30. The mean age of veterans increases between the 1998 and 2000 entry cohort, from 41.6 to 44.8. Veterans are older because a large percentage of them are military retirees, as seen in Figure 2.2. The percentage of veteran entrants who are retirees varies across entry cohorts and ranges from 60 percent in 1994 to 88 percent in 2000. The figure also shows that the majority of these retirees were enlisted at retirement (enlisted retirees) with the percentage of all veteran entrants who are enlisted retirees increasing from 31.7 percent in 1994 to 50.4 percent in 2000. In our sample, these enlisted retirees had at least bachelor's degrees. On average, they accounted for 60 percent of all veteran retiree entrants.

Although mean age has been relatively stable, conditional on veterans' status, the percentage of college-educated GS entrants who are veterans increased dramatically between the 1988 and 2000 entry cohorts, from 5.5 percent in 1991 to 23.5 percent in 1994 and remaining relatively constant thereafter (Figure 2.3, measured by the right axis). Although veterans have been accorded some degree of preference in federal hiring since the Civil War, those preferences were expanded in the 1990s to give additional preference to everyone who served on

Figure 2.1
Mean Age of Entry Cohort, Veterans Versus Nonveterans

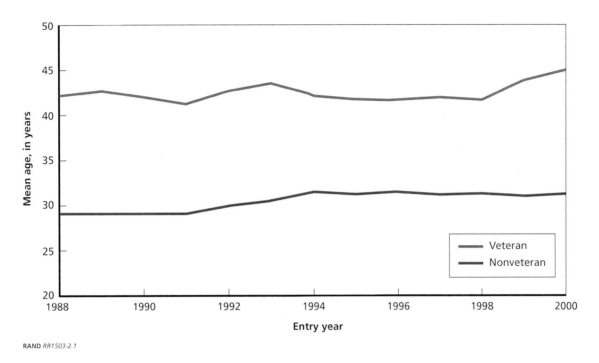

Figure 2.2
Percentage of Veteran Entrants Who Are Military Retirees or Enlisted Retirees

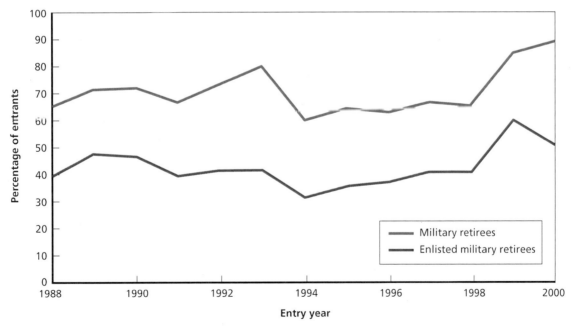

Figure 2.3
Mean Age of Entry Cohort and Percentage Who Are Veterans

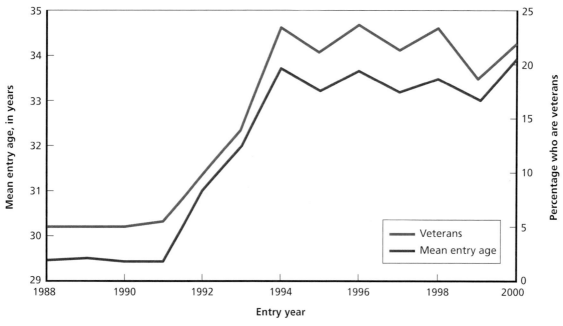

active duty during the first Gulf War, regardless for how long or where.[2] Additional preferences are also given for a compensable service-connected disability and other aspects of service, such as receipt of a campaign medal or Purple Heart. Furthermore, the uniformed services underwent a substantial drawdown in the early 1990s, especially in the Army and Air Force, thereby increasing the pool of veterans searching for employment in the 1990s.

Because the composition of entrants changed over time toward a greater percentage of veterans, and veterans are older, the mean age of entrants increased as well, tracking closely the rise in percentage of entrants who are veterans (Figure 2.3, measured by the left axis). The mean age increased from 29.5 in 1991 to 33.7 in 1994.

Figure 2.4 shows that the grade composition of the entering cohorts also changed over time, shifting toward higher grades. The percentage of employees with bachelor's degrees only entering in grades GS-1 to GS-5 fell from 41.6 percent in 1988 to 31.0 percent in 1992, then to 25.4 percent in 1993, and it remained near that level thereafter. The percentage entering as GS-6 to GS-10 employees increased from 50.7 percent in 1999 to 58 percent in 1992 and reached 62 percent in 2000. The percentage in the highest grades, GS-11 to GS-15, more than doubled, from 7.7 percent in 1988 to 17.2 percent in 1994, and then gradually fell to 14.9 percent in 2000. The tabulation focuses on those with only bachelor's degrees to control for shifts in educational attainment over time.

---

[2]   According to the Office of Personnel Management (OPM), undated (b),

  National Defense Authorization Act for Fiscal Year 1998 (Public Law 105-85) of November 18, 1997, contains a provision (section 1102 of Title XI) which accords Veterans' preference to everyone who served on active duty during the period beginning August 2, 1990, and ending January 2, 1992, provided, of course, the veteran is otherwise eligible.

Figure 2.4
Grade Distribution of GS Entrants with Only Bachelor's Degrees, by Entry Cohort

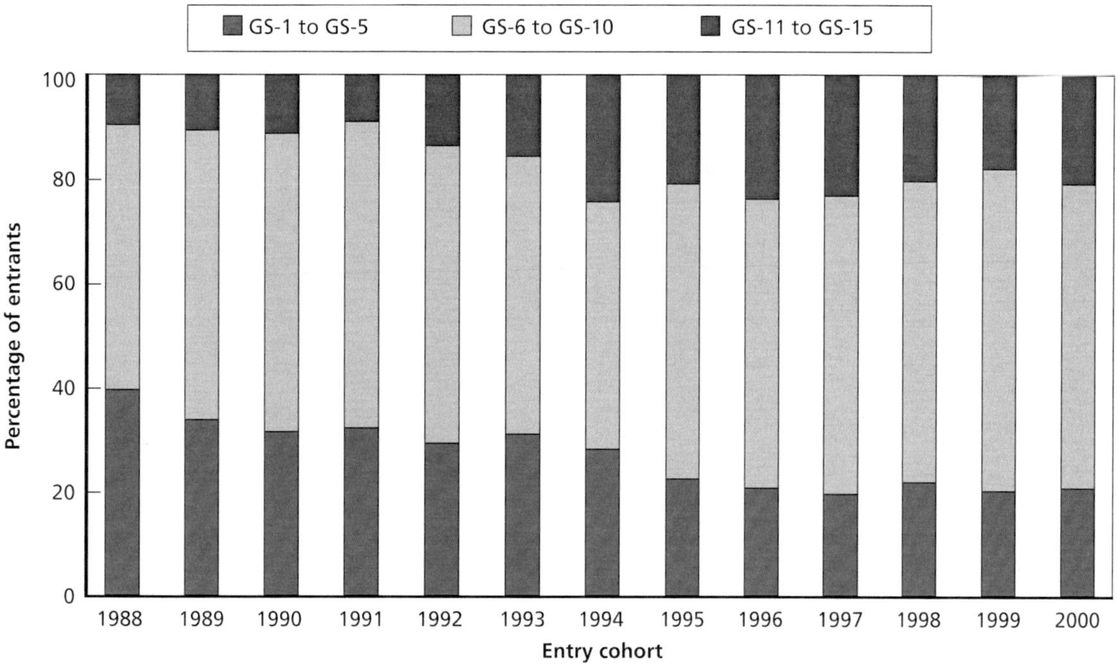

RAND *RR1503-2.4*

The changing grade distribution might in part reflect the greater representation of veterans because veterans enter at higher grades, as shown in Figure 2.5. However, the grade composition also favored higher grades over time among nonveterans (Figure 2.6). The shift toward higher grades even among nonveterans might be a result of the drawdown policies of the 1990s that targeted younger personnel in lower grades (Chu and White, 2000), technological change that is biased toward positions with higher starting pay grades, or a shift in the overall population toward greater educational attainment.

We also observe a shift in the education distribution among GS entrants with bachelor's degrees or higher (Figure 2.7) toward those with greater educational attainment. The educational categories in the figure capture those with given degrees or more but less than any of the higher categories. For example, the category BA includes those with more than bachelor's degrees but without master's degrees or higher. We find that the percentage of entrants with bachelor's degrees decreased from 83 percent in 1988 to 68 percent in 1994 but increased thereafter up to 75 percent in 2000. In contrast, the percentage with master's degrees increased from 13 percent in 1988 to 24 percent in 1994, dropping slightly to 22 percent in 2000.

The shift toward greater education in the early 1990s in part reflects the changing veteran composition and the higher education of veterans (Figure 2.8). However, the shift toward more education in Figure 2.7 began in 1989, before the rise in veterans' representation, suggesting that other factors were at play. Furthermore, in both Figures 2.7 and 2.8, we observe smaller shifts back toward less educational attainment between 1994 and 2000. In the case of veterans, this shift might be due to the increase in the percentage of entrants who are enlisted military retirees. Insofar as enlisted personnel are less likely than officers to attain education beyond bachelor's degrees, the increasing representation of enlisted retirees between 1994 and 2000

Figure 2.5
Grade Distribution of Veteran GS Entrants with Only Bachelor's Degrees, by Entry Cohort

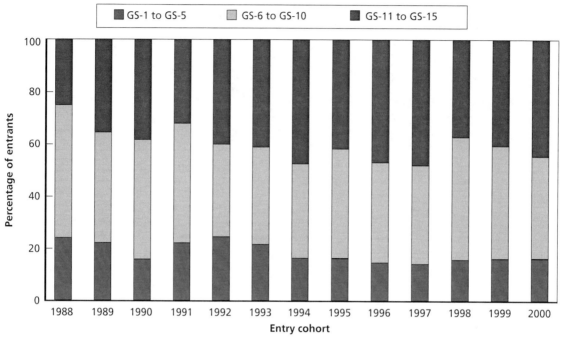

Figure 2.6
Grade Distribution of Nonveteran GS Entrants with Only Bachelor's Degrees, by Entry Cohort

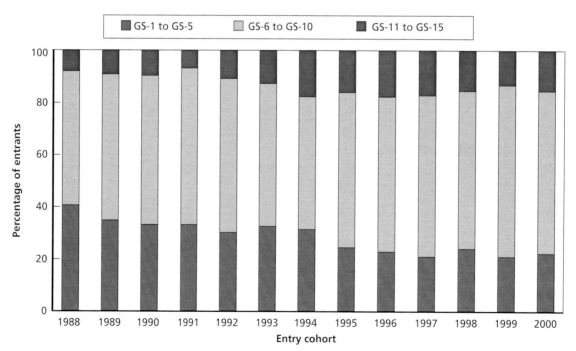

Figure 2.7
Education Distribution of GS Entrants, by Entry Cohort

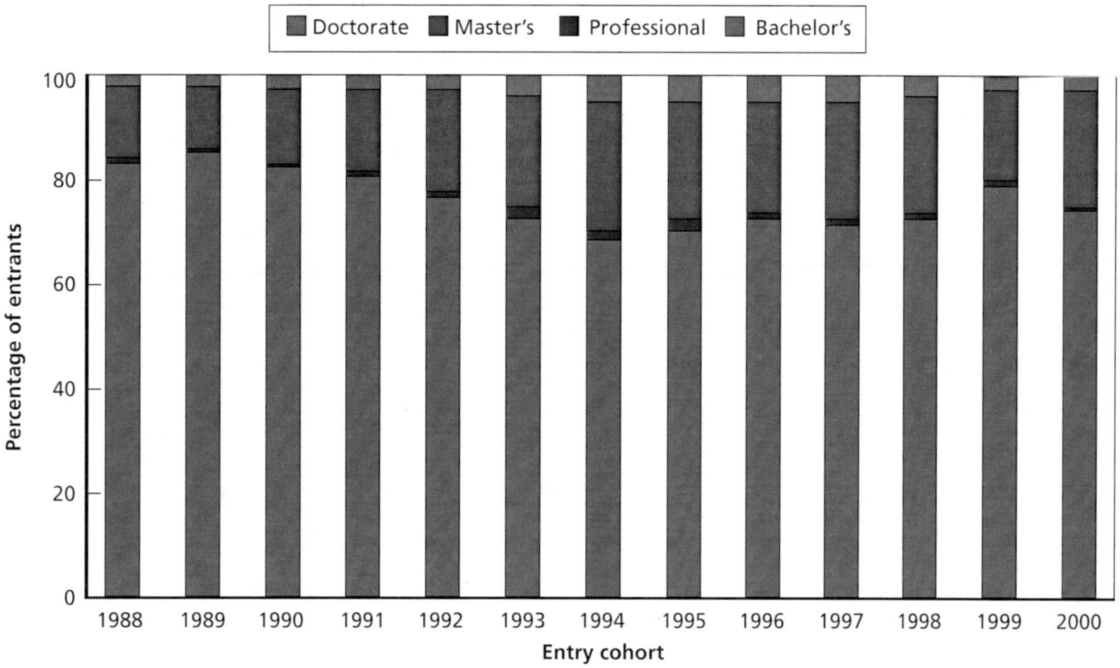

shown in Figure 2.2 might explain this pattern for veterans in Figure 2.8. However, we see a similar pattern for nonveterans (Figure 2.9). The percentage of GS entrants with bachelor's degrees only among nonveterans with at least bachelor's degrees began to drop in 1989 from 86 percent to 70 percent in 1994 but then increased again to 79 percent in 2000. The causes for this post-1994 shift in educational attainment among nonveteran entrants are unclear, although, as we show next, over this same period, the percentage of nonveteran entrants who are female also increased.

Figure 2.10 shows the percentage of GS entrants with bachelor's degrees or more who are female, by veterans' status. The percentage who are female is lower for veteran entrants than nonveteran entrants. Interestingly, the percentage female among veterans increased over time, from a low of 5.8 percent in 1989 to a high of 17.1 percent in 1998, but then ebbed to 13.6 percent in 2000. The changing percentage of females among veteran entrants could reflect such factors as the rising percentage of female officers and enlisted personnel in the military in the 1980s and a difference in male and female separation rates during defense drawdown. For nonveterans, the percentage of entrants who are female fluctuated: 43.0 percent in 1988, 53.6 percent in 1993, 44.6 percent in 1996, and 49.9 percent in 2000.

Finally, we consider the percentage of entrants who are minorities (Figure 2.11). This percentage was higher for nonveteran entrants than for veteran entrants from 1988 to 1998, but the percentages were nearly equal in 1999 and 2000 at around 25 percent. In particular, the percentage rose faster for veterans, bringing it to the level for nonveterans.

Figure 2.8
Education Distribution of Veteran GS Entrants, by Entry Cohort

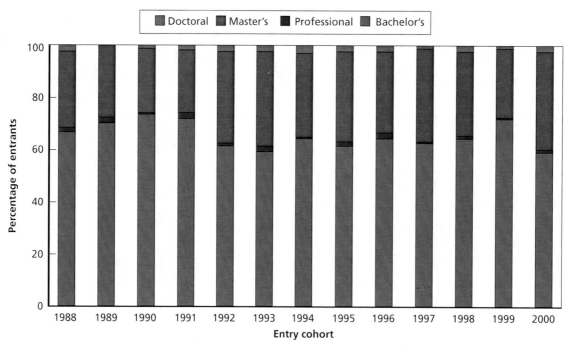

RAND RR1503-2.8

Figure 2.9
Education Distribution of Nonveteran GS Entrants, by Entry Cohort

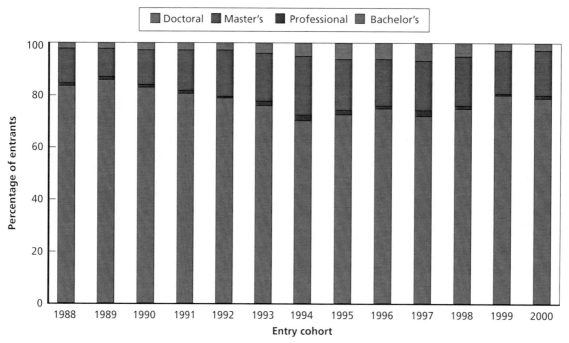

RAND RR1503-2.9

Figure 2.10
Percentage of GS Entrants with At Least Bachelor's Degrees Who Are Female

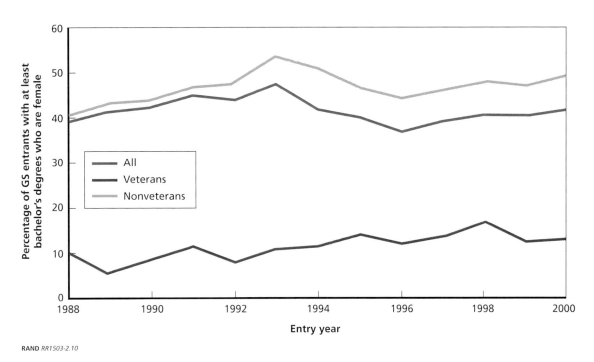

RAND RR1503-2.10

Figure 2.11
Percentage of GS Entrants with At Least Bachelor's Degrees Who Are Minorities

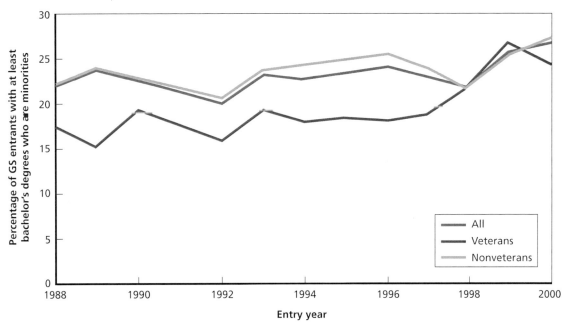

RAND RR1503-2.11

Entry Cohort Retention

Figure 2.12 presents the average retention rates among our sample of DoD civil service workers entering between 1988 and 2000. We include both 1988 and 2000 retention rates separately to demonstrate that entry cohort retention has varied over time. Roughly 25 percent of the sample exits the DoD civil service by the third year of employment, although, for entrants in 2000, this threshold was not crossed until the fourth year of service. The difference between the 1988 and 2000 entry cohort increased over time: Sixty percent of the original 1988 entry cohort was retained after seven years, while the 2000 entry cohort reached 60-percent retention after ten years of service.

The differences in entry cohort retention rates could be driven in part by sample composition, or it could also represent DoD agency or calendar time characteristics that we do not explicitly model, such as periods of substantial downsizing. We examine compositional effects in Chapter Five and discuss how agency-time characteristics might affect both our model estimates and our policy simulations.

Summary

Among our sample of GS entrants with at least at bachelor's degrees, we found demographic differences across entry cohorts. Probably the most notable change was the increase in the percentage of entrants who are veterans. With this change came an increase in the mean age of entrants, given that veterans are older and a majority of veterans are military retirees. We found a shift toward higher entry grades, controlling for educational attainment, across the 1988-to-

Figure 2.12
Retention Rates Among U.S. Department of Defense Civil Service Entrants from 1988 to 2000 with At Least Bachelor's Degrees

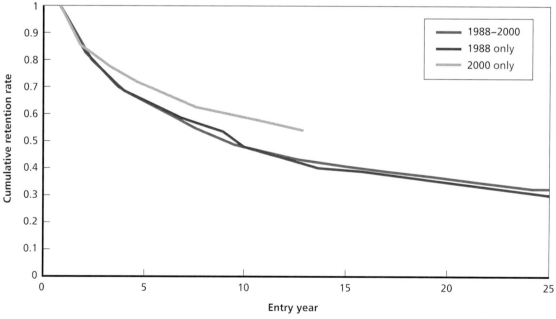

2000 entry cohorts, among both veteran and nonveteran entrants and a shift toward higher levels of education overall. We found a rising percentage of employees who are female and who are minorities, among both veteran and nonveteran entrants. In the next chapter, we discuss the implications of the demographic differences and changes across entry cohorts.

CHAPTER THREE

The Dynamic Retention Model: Overview and Extensions

The research summarized in this report extends the DRM for DoD civil service personnel by considering group differences in tastes for military service within an entry cohort and estimating models for additional cohorts. These extensions permit us to determine whether the parameter estimates and, consequently, the simulated effects of policy changes are sensitive to differences across entry cohorts. It also extends the DRM simulation capability by including changes in total personnel costs in addition to changes in retention resulting from changes in compensation, and it uses more-refined pay profiles to represent the civil service and external pay opportunities over a career. Chapter Four discusses the pay profiles. Here, we discuss the extensions regarding group and entry cohort differences, as well as our costing methodology. We begin with an overview of the DRM for civil service employees. This draws from our earlier document (Asch, Mattock, and Hosek, 2014a). That document provides a review of the earlier literature.

Overview of the Dynamic Retention Model for Civil Service Employees

The DRM is an econometric model of retention behavior. In it, each employee makes a retention decision each year of whether or not to continue their career with a given employer. The model assumes that the employees are rational and forward looking, taking into account their expected future earnings from the employer, their own preference for employment with that employer, and uncertainty about future events that could cause them to value their current service more or less than their external opportunities. Once we estimate the parameters of the underlying decision process, described below, we can use these estimates to simulate the retention profile for an entry cohort of civil service personnel under the current compensation policy and alternative policies of interest. By appropriately scaling the results, we can make inferences about those policies' effects on the size of the workforce that is retained and the accessions needed to sustain the size should it decrease.[1]

We model civil service retention from the start of an employee's career in the civil service. Employees in the model can enter the civil service for the first time at any age. Our sample is limited to entrants with four or more years of college and between the ages of 25 and 52; the bulk of entrants are in this age range.

---

[1] We can infer accession requirements if the workforce should decrease under a policy. However, the model and data do not extend to the accession decision, i.e., the individual's choice to enter the DoD civil service.

Each year in the model, the individual employee is assumed to compare the value of staying in civil service with that of leaving, and they base the retention decision on which alternative has the maximum value. In the DRM, we assume that once individuals leave DoD civil service, they do not reenter at a later date.[2] Individuals who stay can revisit the stay/leave choice in each future period over an assumed 40-year time horizon corresponding to the period between ages 25 and 65. All of these decisions will depend on the employee's unique circumstances in each period. In addition to civil service and external pay, those circumstances include preference for DoD civil service relative to external opportunities and random events that might affect relative preferences.

In the model, we specify the value of staying in the DoD civil service as

$$V_t^S = \gamma^c + w_t^c + \beta E_t \left[Max\left(V_{t+1}^S, V_{t+1}^L\right)\right] + \varepsilon_t^c,$$

(3.1)

where

V_t^S is the value of staying in the DoD civil service at time t

γ^c is individual taste for DoD civil service relative to the external market

w_t^c is civil service annual earnings at time t

β is the civil service employee's personal discount factor

V_{t+1}^S is the value of staying in DoD civil service at time $t + 1$

V_{t+1}^L is the value of leaving DoD civil service at time $t + 1$, defined in Equation 3.2

$E_t \left[Max\left(V_{t+1}^S, V_{t+1}^L\right)\right]$ is the expected value of having the option to choose to stay or leave in the next period

ε_t^c is the random utility shock to DoD civil service employment at time t.

The value of staying depends on the annual civil service earnings in each time period, w_t^c. Annual earnings vary with age, and those who begin their civil service career at older ages also begin at higher pay than their younger counterparts.[3] As we discuss in Chapter Four, we use a single curve to represent GS salary and another single curve to represent external salary by age.

The value of staying also depends on the individual's preference for DoD civil service relative to the external market (their taste for civil service), given by γ^c. We assume that the distribution of taste among entrants to the DoD civil service is normally distributed with mean μ and standard deviation (SD) σ. The realized taste for civil service of an entrant, γ^c, is assumed to be constant over time for that individual and can be thought of as the net effect of idiosyncratic, persistent differences related to the individual's perceived value of working in the civil service relative to the external market. The net effect includes all nonmonetary and monetary factors the individual perceives as relevant to the civil service over and above monetary factors included in the model. These factors might include patriotism and desire for public service, positive and negative aspects the individual perceives about the civil service (e.g., hours of work, differences in paid leave), and persistent differences in civil service and private-

---

[2] This, in fact, is not true. Civil service employees can flow in and out of DoD civil service. Furthermore, the DRM can accommodate such flows, like we did in Asch, Hosek, and Mattock, 2013, and earlier studies that permit flows of military personnel in and out of the reserve components. In the data we use, 10 percent of full-time, GS entrants with at least bachelor's degrees and enrolled in FERS ever return between entry and the end date of our data, 2012. We therefore exclude this possibility (and these observations) for the purposes of this study.

[3] Intuitively, the wage at entry into the civil service among those who enter can be expected to increase with age because entrants have the alternative of working in the economy at large, in which pay tends to increase with age.

sector earnings apart from the differences accounted for in the expected-pay profiles used in estimating the model. As mentioned, we use single curves to represent GS salary and external salary by age. But someone might believe that their GS and external salaries are persistently higher or lower than those curves. The net effect of these perceived differences would enter into taste. Another way of describing individual taste, then, is as a person-specific fixed effect. This individual fixed effect is known to the individual but not to the researcher. We use the data, together with the structure of the model, to estimate the mean and variance of the taste distribution across individuals, together with the other parameters of the model.

Tastes for civil service differ across civil service employees. We as analysts do not directly observe these tastes, but we assume that they are distributed according to a specific probability distribution. The parameters of this probability distribution of taste, however, are unknown, and a goal of the estimation process is to estimate these parameters.

The decision of whether to stay or leave in Equation 3.1 depends also on a period- and individual-specific environmental disturbance term (or shock) that can either positively or negatively affect the value placed on civil service in a single period, given by ε_t^c. For example, having an ailing family member who requires assistance with home care could be such a shock that could decrease the value placed on civil service employment.

The value of staying in the civil service also includes the value of the option to leave at a later date, given by $\beta E_t \left[Max \left(V_{t+1}^S, V_{t+1}^L \right) \right]$, where V_{t+1}^S and V_{t+1}^L are the value of staying and leaving the next period, respectively, and β is the personal discount factor. The individual knows that they can revisit and reoptimize the decision to stay or leave in the next period. Of course, the future is uncertain, so the value of being able to choose to stay or leave in the future is expressed as the discounted present value of an expected value. Individuals can reoptimize and might change their decisions in the future because new information, e.g., a new shock, makes doing so reasonable or because the discounted expected value of future benefits of leaving becomes greater than the benefits of staying. Furthermore, choices made today can affect the value of choices in the future. Someone who chooses to stay in the civil service today adds a year of service, moving closer to retirement eligibility and increasing retirement benefits, thereby influencing the value of choosing the civil service in the future. Similarly, past choices can affect the value of current and future choices.

The value of leaving DoD civil service at time t is

$$V_t^L = w_t^e + \Sigma_{s=t+1}^T \beta^{s-t} w_s^e + R_t^c + \varepsilon_t^e,$$

(3.2)

where
   V_t^L is the value of leaving DoD civil service at time t
   w_t^e is annual earnings in the external market at time t plus retirement benefits that will accrue to the civil service employee in the external market from t until T
   $\Sigma_{s=t+1}^T \beta^{s-t} w_s^e$ is the present value of future external market earnings
   R_t^c is the present discounted value of the retirement benefit accrued as a result of civil service employment for someone leaving at time t (exclusive of any benefit accrued from work in the external market)
   ε_t^e is the random shock to external employment at time t.
   The value of leaving includes the value of the external alternative, which includes the present value of pay in the external market, given by w_t^e; plus the discounted sum of future

pay; any civil service retirement benefits the individual is entitled to receive (R_t^c); and an individual- and period-specific shock term that can either positively or negatively affect the value of the external alternative (ε_t^e). Pay in the external market varies with age, with those entering the civil service at older ages having higher external pay opportunities. Entry age also affects civil service retirement benefits.

Someone who leaves DoD civil service might either move to another federal agency or leave federal service altogether. Thus, the value of leaving could include federal pay and expected federal retirement benefits in the next federal job. We could extend the DRM to incorporate movement to other federal agencies. However, with the available data, we do not know how long people stay in federal employment if they are not in DoD. Consequently, in the current analysis, we assume that all DoD separations are separations from federal employment, although we recognize that this leads to measurement error of the value of external opportunities for those who transfer.[4]

We assume that, to claim civil service retirement benefits, the individual must have left the civil service. That is, we do not include the possibility that a civil service employee might become a reemployed annuitant and claim retirement benefits while working in the civil service. We excluded this possibility to simplify our empirical analysis, although the DRM can be extended to allow the choice to become a reemployed annuitant. FERS covers all the DoD civilians in our data. It consists of three parts: a defined benefit plan called the Basic Benefit Plan that bases retirement benefits on the employee's earnings and years of service, Social Security coverage, and a defined contribution plan (the TSP). All new entrants with less than five years of service are automatically placed under FERS.[5] Regarding Social Security, we assume that civil service employees who separate then take employment that Social Security also covers. Thus, when computing the expected retirement benefit under FERS, we net out the expected Social Security retirement benefits. Private-sector retirement benefits in the form of defined contribution plans are absorbed into the taste term as a persistent difference in compensation. In this framework, we do not capture private-sector defined benefit plans because each plan follows a unique accrual plan. These plans have become significantly less common

---

[4]   Missing the benefit of continued FERS pension benefit growth in the non-DoD civil service job causes measurement error in the value of external opportunities. This would cause the parameter estimates of taste (see Chapter Five) for DoD civil service to be biased downward because they would miss the option value associated with the transition from DoD civil service to non-DoD civil service. This bias would cause our responses to policy changes to be larger than if the model were estimated accounting for DoD-to–non-DoD civil service transitions.

[5]   The TSP portion of FERS includes agency-matching contributions. We do not model employees' decisions on how much to contribute to the TSP and therefore the amount of the DoD match. Available data on TSP participation from the Federal Retirement Thrift Investment Board indicates that between 80 and 90 percent of federal employees participate in the TSP, with an average contribution rate exceeding 8 percent between 2009 and 2013. The 8-percent rates imply an agency match rate of 5 percent (Federal Retirement Thrift Investment Board, undated). We assume a 5-percent match rate in our DRM estimation. We also assume a 5-percent rate in our simulations except for the cases in which we explicitly consider lower rates, as discussed in Chapter Seven. We also assume an annual real TSP return of 5 percent. Rates of return vary significantly depending on the fund in which the individual chooses to invest. The two most popular funds as of December 2013 for someone with at least a bachelor's degree, the Government Securities Investment Fund and Common Stock Index Investment Fund, have had a nominal annual rate of return since inception equal to 5.29 percent (April 1987) and 10.03 percent (January 1988), respectively. The average inflation rate, as measured by the Consumer Price Index for All Urban Consumers, for the same time period was 2.5 percent, suggesting that 5 percent is a reasonable average real rate of return for our sample (TSP, undated [a]; TSP, undated [b]).

in the private sector, with only 11 percent of private-industry employees participating in such a plan in 2011 (Wiatrowski, 2012).

Structural models, such as the DRM, have been used in other work. For instance, dynamic programming has been applied to analyze retirement decisions and choices between full- and part-time work (van der Klaauw and Wolpin, 2008). Such models use a period-specific utility function, and the objective is to maximize intertemporal utility subject to initial assets, saving behavior, and the retirement system, e.g., Social Security. Such specifications are potentially useful for analyzing civil service retention or military retention, but available data limit what can be done. Data on spouse earnings, full- versus part-time work, savings, wealth, and the timing of retirement are absent, for example. Our value-function specification can be thought of as a particular form of utility function in which current utility depends additively on the current wage, taste, and shock, plus the discounted expected value of following the best path in the next period.

We do not observe individuals' tastes for the civil service or random shock terms. Instead, we assume they are each distributed according to a known probability distribution with unknown parameters that we estimate using available data. Specifically, we assume that individuals' tastes for civil service are normally distributed and that the random shocks have an extreme-value type 1 distribution. Given these distributional assumptions, we can derive choice probabilities for each alternative at each decision-year and the cumulative choice probabilities or survival probabilities for an entering entry cohort at each decision-year and then write an appropriate likelihood equation to estimate the parameters of the model. Our earlier document (Asch, Mattock, and Hosek, 2014a) writes out the choice probabilities, derives the expression for the likelihood function, and discusses our use of maximum likelihood to estimate the model. Those expressions and that discussion are equally relevant to the analysis in this study.

The parameters we estimated are the mean and SD of the taste distribution, the scale parameter of the shock distribution, and the discount factor. Like we discuss in our earlier report, we estimated two additional parameters: the probability of attrition in the first year of DoD civil service employment and the probability of being censored in 1997, when reorganization resulted in some DoD employees being moved to an organization outside of DoD.[6] When we estimate the DRM by entry cohort, we include the second additional parameter related to censoring only for cohorts entering prior to 1996. In addition to these six parameters, we estimate parameters related to group differences—specifically, veterans' status—and, in models in which we combine entry cohorts, we allow some parameters (e.g., the mean and SD of tastes) to vary across cohorts. In the next section, we discuss the DRM extensions related to group differences and cohort differences.

To judge goodness of fit, we used the parameter estimates to simulate retention by year of service for DoD civil service personnel; aggregate individual retention; and compare the aggregated retention rates, the rates, and the actual data. We show goodness-of-fit diagrams, including error bands, around our model predictions in Chapter Five when we present the model parameter estimates.

---

[6] The reorganization in 1997 was that some DoD civil service members—notably, cartographers—were moved to the newly created National Imagery and Mapping Agency. Consequently, personnel in some occupations disappeared from (were censored from) the DoD civil service in 1997.

We use simulation to consider the retention and cost effects of changing civil service compensation policy. In the policy excursions in Chapter Seven, we simulated the cumulative retention probabilities at each year of service under baseline conditions (i.e., the policy status quo, without any policy changes) and then under the considered policy change. We show simulated effects of policy changes in the steady state, meaning after all individuals in the civil service have spent their entire careers under the new policy regime. The policy excursions focus on changes to the FERS defined benefit plan that Congress recently passed. We discuss these excursions in more detail in Chapter Seven. We rescaled the profiles to the fiscal year 2011 end strength. At the end of fiscal year 2011, the end strength for the DoD GS workforce with bachelor's degrees or more was 225,888. This rescaling allows us to present the simulation results in terms of the size of the workforce. An important extension of the DRM capability in this study was to expand the simulation coding in the model to assess cost, as well as retention effects, of policy changes. Our earlier work considered only retention effects. In the next section, we discuss how we simulated cost.

Dynamic Retention Model Extensions

The DRM in Asch, Mattock, and Hosek, 2014a, used data on the 1988 entry cohort of DoD GS civilian employees and estimated the six-parameter model described earlier. We extend the DRM by using data on entry cohorts for 1988 to 2000 and by considering the possibility that veterans have a different distribution of tastes for DoD civil service from that of nonveterans. That is, we estimate the DRM with additional entry cohorts and allow for group differences in tastes within an entry cohort and across entry cohorts.

Taste Distribution Differences Across Entry Cohorts and Across Groups

In the DRM used in Asch, Mattock, and Hosek, 2014a, we estimated the mean and SD of the taste distribution of the 1988 entry cohort. Together with the other parameter estimates, we showed that the model fit is extremely good, and, in fact, we conducted out-of-sample predictions and found that the model performs very well. On the basis of these estimates, we simulated the retention effects of the 2011–2013 pay freeze, as well as furloughs.

A natural question is whether the experience and retention behavior of the 1988 entry cohort are representative and whether we would continue to find similar estimates and retention responses to policy changes using data for other cohorts. At the heart of this question is whether the taste and shock distribution parameters differ across entry cohorts. We focus here on the taste distribution.

There are two reasons that the mean and SD of the taste distribution might change across entry cohorts. The first is that individual tastes can change over time. For example, it is common for the popular press to point out differences in the career aspirations and cultural attitudes of generation X (born between 1965 and 1980, according to the Pew Research Center), millennials (born between 1980 and 2000), baby boom (born between 1945 and 1965), and Depression-era and wartime (born between 1925 and 1945) generations. The common argument is that younger generations evaluate their labor market opportunities and establish priorities that differ from those of older generations. However, Stafford and Griffis, 2008, reviews available data and literature on how the millennial generation differs from earlier generations and argues that some of the characteristics attributed to millennials or, for that matter, any specific generation at a point in time, might be due to life-stage effects that are found in all generations as they age.[7] That is, different generations might respond in a similar fashion when they are at the same age. An alternative view is that calendar year, not birth year, is the most relevant for decisionmaking. For example, one could argue that changes in societal attitudes and culture over time affect all individuals in a similar fashion, regardless of birth year. Thus, the tastes of all individuals entering public service in 2000 might differ from those entering in 1988, regardless of age.

Ultimately, whether the taste distribution has changed over time across entry cohorts is an empirical issue. We investigate this issue by estimating separate DRMs for each entry cohort. We then estimate a combined model using data for all entry cohorts that includes a separate mean taste parameter for each entry calendar year. We estimate the combined model for a specific demographic subset of the data to limit the possibility that the observed differences are due to changes in the demographic composition of the entry cohorts over time. In particular, we estimate a combined model for nonveterans, ages 30 or younger, with bachelor's degrees and no further education. We then test the statistical significance of the mean taste parameters across years for evidence of a shift in tastes across entry cohorts, holding group composition constant.

As we show in Chapter Five, point estimates of mean tastes differ across entry cohorts but are not always statistically significant when we hold group composition constant as mentioned. The responsiveness to a pay change also differs across entry cohorts, but the differences are not large.

A second reason that the estimated taste distribution parameters might differ across entry cohort is that entry cohort composition changes over time. Chapter Two presented tabulations of the demographic characteristics of the entry cohorts and how they changed over time. The most dramatic change was in the percentage of entrants who were veterans, increasing from 5.5 percent in 1991 to 23.5 percent in 1994. This increase accounted for much of the increase in mean age over the same period. The changes in veteran representation also affected changes in the grade distribution, education distribution, and gender distribution, although these changes also occurred among nonveteran entrants.

Although demographic differences do not automatically translate into differences in tastes for DoD civil service, the distribution of tastes for DoD service could differ between veterans and nonveterans, especially between veterans who are military retirees and nonveterans. Military retirees have already spent at least 20 years working in support of the military mission and have thereby revealed a high taste for DoD employment. Indeed, our DRM estimates for military personnel show that mean taste increases with years of service as a result of selective retention: Lower-taste personnel separate and higher-taste personnel stay, while the SD of tastes decreases because personnel who stay have more-homogeneous tastes (Mattock, Asch, et al., 2014). Consequently, we test whether military veterans have a different taste distribution from that of nonveterans and specifically have a higher mean taste and lower SD of taste. We estimate the mean and SD of the taste distribution for nonveterans and include parameters that allow the mean and SD of the taste distribution for veterans to differ linearly. If these additional parameters are statistically significantly different from 0, the mean and SD differ

---

[7]   None of the entrants in our data are millennials because the youngest person is age 22 in 2000, born in 1978. As the tabulations of mean age show in Figure 2.1 in Chapter Two, nonveterans include baby boomers and generation X-ers, while the veterans include mostly baby boomers.

between veterans and nonveterans. We also allow the mean and SD of tastes to shift across entry cohorts, separately for veterans and nonveterans, in our combined models and entry cohort–specific models.

Ideally, we would want to consider adding parameters that allow the taste distribution to shift for different demographic groups. This would allow us to test whether the distribution differs across groups (e.g., boomers and generation X-ers). However, adding parameters also increases the computational burden of estimating the model. For this reason, we focus on differences in the taste distributions between veterans and nonveterans.

We present and discuss the model estimates in Chapter Five.

Costing Methodology

In addition to simulating retention under baseline and alternative policies, we compute the associated cost of each policy. We compute the personnel costs associated with current compensation and those associated with retirement and sum them to compute the total cost of the baseline force and the force under the policy change. Current compensation costs are the costs of federal civil service pay for the retained force. Although we do not analyze the retention and cost effects of offering retention bonuses in Chapter Seven, we could do so, and current compensation costs would also include bonus costs.

Retirement costs are the cost of FERS. TSP costs are the costs of agency contributions, assumed for simplicity to be 5 percent for all employees.[8] The Federal Retirement Thrift Investment Board retains the agency's contributions to an employee's TSP account if the employee leaves before vesting and are, therefore, still a cost to the agency. To compute the cost of the Basic Benefit Plan, we compute the normal cost percentage of the pay bill for the force, consistently with OPM's actuarial practice. This gives an amount—an accrual charge—sufficient to cover the retirement liability of the workforce that retires from DoD civilian service under FERS. This amount is computed based on the civil service's distribution of work history projected from a specific policy (the policy could be the baseline policy or an alternative). The accrual charge is the cost to U.S. taxpayers of providing the Basic Benefit Plan to the civil service workforce. However, because employees are required to make contributions to help fund their basic plans, the cost to DoD is the accrual charge minus the share of the accrual charge that employee contributions cover. Employees first hired before 2013 contributed 0.8 percent of pay toward the basic plan. Those hired in 2013 are required to contribute 3.1 percent, while those hired after 2013 are required to contribute 4.4 percent. In our baseline simulations, we assume a 0.8-percent contribution rate for employees, and, in Chapter Six, we show simulations assuming the higher 4.4 percent. We compute retirement costs in two ways: the cost of the retirement plan (the full accrual charge) and the effect of the retirement plan's cost on DoD (the accrual charge minus employee contributions). The cost of the retirement plan is the full cost of providing the benefit, which DoD and the employees bear. Total costs to DoD are the sum of current costs and retirement costs that DoD bears.

---

[8]   We do not estimate a distribution of contribution rates, but doing so might be a fruitful path for future research.

CHAPTER FOUR

Estimates of Federal and Private-Sector Pay Profiles

An individual's decision to stay in the DoD federal civil service or leave and enter the external market relies, in part, on current and deferred compensation. For the DRM to accurately reflect the monetary incentives underlying this decision, we model what someone can expect to earn from working in the federal civil service versus working in the private sector. A pay profile reflects someone's potential earnings based on age and sector. The emphasis is on potential earnings. We abstract from job search and unemployment and assume that civil service workers can find an outside job offer and face no risk of job loss. We estimate pay profiles using regression analysis to control for age, gender, education, and veteran status.

We make several methodological improvements over earlier versions of the civil service DRM. First, we account for top-coding of the pay data, which is particularly important in the private sector for highly educated individuals older than 40. Top-coding is common in publicly available data sets as a means of protecting individuals' privacy. Top-coding has the potential to bias the estimated pay profiles downward because earnings above the top-code threshold are not reported but are merely flagged as top-coded. If not controlled, the estimated earning increase with age will be biased downward, which then can distort incentives in the DRM to stay or leave the civil service. Second, because we are estimating the DRM over many entry cohorts, we were concerned that differences might exist in pay profiles across birth cohorts. To explore this possibility, we created synthetic birth cohorts using repeated cross-sectional data from the CPS for 1966 through 2014 (U.S. Census Bureau, undated) and compared the pay profiles from these synthetic birth cohorts with pay profiles calculated using only cross-sectional information. After accounting for differences in year, gender, and educational composition of the birth cohorts, we find only small differences in pay profiles by birth cohort.

After accounting for top-coding, we find the pay profiles estimated from the ACS for 2003 through 2012 to be very similar to pay profiles estimated from the CPS for the same years. In light of the larger sample in the ACS and the similarity between the pay profiles estimated using synthetic birth cohorts versus cross-sectional data, we use the simpler cross-sectional pay profiles from the ACS in our final analysis. We describe our pay analysis in what follows.

Overview of the Data

We used two nationally representative data sets to estimate pay profiles: the CPS Annual Social and Economic Supplement (U.S. Census Bureau, 2015) and the ACS. Each data source has its advantages. The CPS has been conducted from 1962 until the present day, collecting detailed information on household earnings, demographics, and more. However, the sample is small, with 106,693 observations between ages 25 and 65 in March 2012, of which only 1,328 are federal employees with at least bachelor's degrees who are working full time. The ACS has been conducted only from 2000 until the present day but with a much larger sample. In 2012, the ACS interviewed 1,662,892 individuals between the ages of 25 and 65, including 14,983 federal employees with at least bachelor's degrees and working full time.

We estimate pay profiles separately for private and federal civil service employees. To be included in either sample, an individual had to be male, work full time over the past year (usually work 35 or more hours per week, 40 or more weeks per year), report positive earnings in the past year, have at least an undergraduate degree, and be between the ages of 25 and 65.[1] We identified an individual as working in the federal civil service if they reported being a wage or salaried employee for the federal government and not in the armed services. We identified an individual as working in the private sector if they reported being a wage or salaried employee for a private employer. For our estimates of pay profiles based on recent cross-sectional data, we consider years 2003 through 2012, which were available for the ACS at the time of the analysis. When estimating pay profiles for individuals by birth cohort, we used the CPS for 1966 through 2014, and we further restricted the sample to those born between 1941 and 1985. We converted all dollar values to 2013 dollars using the Consumer Price Index for All Urban Consumers published by the Bureau of Labor Statistics.

The primary variable of interest is the individual's self-reported pretax wage and salary income. This includes wages, salaries, commissions, cash bonuses, tips, and other money income received from an employer. Figure 4.1 presents the observed mean and median of wage and salary income from the cross-section of federal and private workers in the ACS sample. The figure shows that (1) average income rises until the mid-40s; (2) median federal civil service and private earnings are similar from ages 35 to 55; and (3) the difference between mean and median earnings is large for the private sector, implying greater variance in the earning distribution for workers in the private sector. In part, this results from the cap on federal civil service salaries on the GS (the most common federal pay schedule). The cap was $155,500 in 2013, whereas private-sector workers have no earning cap.

Neither the CPS nor the ACS tracks the same individuals across years. However, CPS data can be combined to create synthetic birth cohorts by tracking the earnings of an average individual in our sample. For example, consider full-time federal civil service workers born in 1943. We can observe the average earnings at age 25 in 1968, age 30 in 1973, age 35 in 1978, and so forth until age 65 in 2008. Because of the smallness of the CPS sample, we grouped individuals in birth cohorts that included observations two years to either side. For example, the 1943 birth cohort included individuals born between 1941 and 1945.

---

[1]   Our restriction to only men might be controversial. In the case of federal civil service workers, we want our sample to reflect the population of individuals who, having maintained a continual attachment to the labor force, are currently employed by the federal civil service. The DRM is a life-cycle model, meaning that it follows individuals from the time they enter the federal civil service until the time they exit. As such, the focus is on career workers. In the CPS and ACS, we do not observe total years someone has been in the labor force, so our sample restrictions are intended to limit the sample to likely career workers. Not including women means that our pay profiles abstract away from possible gender discrimination that differentially affects federal civil service earnings and private-sector earnings for female employees versus male employees. To the degree that gender discrimination has such a permanent differential effect on earnings, we could accommodate its inclusion by allowing taste for federal civil service to differ by gender (although discrimination would not be directly identified because of other permanent, concurrent differences that could also explain differential retention, such as nonmonetary benefits—work hours and work–life balance). This could be an interesting extension for future research.

Figure 4.1
Male Wage and Salary Income, by Age, in the American Community Survey for 2003 Through 2012 for Full-Time, Full-Year Private and Federal Employees

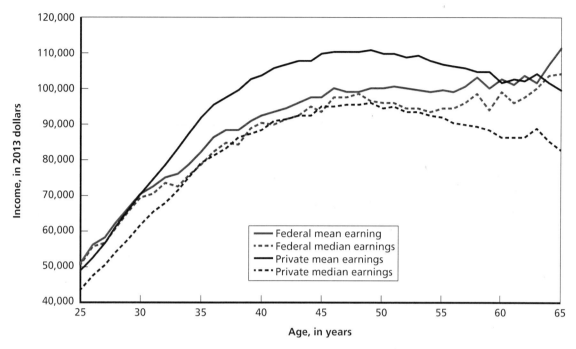

RAND *RR1503-4.1*

Figure 4.2 demonstrates the average pay profiles for the 1953 and 1973 birth cohorts. We observe that federal civil service workers earn less on average than private-sector workers. Additionally, for both sectors, the 1973 birth cohort has a steeper earning trajectory through the early 40s. The patterns in Figure 4.2 might reflect top-coding, differences in sample composition across time (e.g., more–highly educated workers might not enter the job market until later ages), or year effects. Alternatively, perhaps earlier generations had less steep earning trajectories, which is something that might alter their stay/leave decisions. Later in this chapter, we propose a method to control for observable factors that might explain why the pay profiles differ by birth cohort.[2]

Top-Coding

Top-coding is the censoring of earnings above a threshold. It protects the confidentiality of individuals with high incomes who might otherwise be identifiable if income were reported along with such information as state, occupation, marital status, age, and education.

---

[2]   The focus of our analysis is not on how federal and private-sector pay levels compare, and a review of that literature is beyond the scope of our analysis. A review and a recent comparison of federal and private-sector wages can be found in Falk, 2012. That said, the results in Figure 4.1 are qualitatively similar to those in the Falk study; that study found that private-sector wages are more dispersed than federal wages and that federal employees with bachelor's degrees have about the same wages as private-sector workers with similar observable characteristics.

Figure 4.2
Male Average Wage and Salary Income, by Age and Birth Cohort, in the Current Population Survey for 1962 Through 2014 for Full-Time, Full-Year Private and Federal Employees

RAND *RR1503-4.2*

The top-coding threshold varies by survey, year, and income variable. Since 2003, the ACS has top-coded salary and wage income above the 99.5th percentile within each state. For example, in 2012, Idaho had the lowest threshold at $150,000, while the District of Columbia had the highest threshold at $375,000. We used the 2012 top-coding thresholds, rescaled to 2013 dollars, to identify observations that were top-coded.

The CPS changed how salary and wage income were reported in 1988. Instead of a top-coding threshold for salary and wage income from all jobs, the Bureau of Labor Statistics used wage and salary income from the longest job and from other jobs. A top-coding threshold was provided for each variable. In 2012, the threshold was $250,000 for the longest job and $50,000 for other wage and salary income.

Our interest is the distribution and average pay of federal civil service workers and private-sector workers. Although it is a necessity for privacy, top-coding will bias pay profiles lower if we calculates them using these data. For example, suppose half of a sample had earnings of $100,000, and the other half had earnings of $200,000. The sample average should be $150,000. However, if the data are top-coded at $180,000, the sample average is only $140,000. As more of the sample receives pay that exceeds the top-coding threshold, the bias becomes greater. As a result, we would expect the bias to be larger for private-sector workers at older ages.

We account for top-coding by estimating a Tobit model (Tobin, 1958). A Tobit model nests a continuous outcome model (i.e., pay conditional on working) inside a binary outcome model (i.e., pay is above or below the top-coding threshold). Assuming that the unobserved residual of the model is distributed normally, the nested model can be estimated using maximum likelihood and produces consistent and unbiased predictions of mean outcomes. Intuitively, a Tobit model exploits the structure of the residual's distribution to correct the mean of the distribution. It does this by redistributing, by the shape of the normal distribution's right tail, the fraction of the sample's pay distribution that is affected by top-coding.

Overview of the Regression Approach

Our Tobit model includes age, education level (bachelor's, master's, professional, or doctorate), veteran status, and year effects. To account for the nonlinear nature of earning profiles, we allow for a linear age spline in five-year increments starting at age 25. When the estimating the model with ACS data, we placed states with similar top-coding thresholds into groups (tiers) and included an indicator variable for each tier. We predict the pay profile for workers who were male, had only bachelor's degrees, and were not veterans, for year 2011. Appendix A provides additional details and regression results. Figure 4.3 presents the predicted pay profile from the models estimated with ACS and CPS data for years 2003 to 2012.

The pay profiles from both models are similar for private and for federal civil service workers. Also, the same earning pattern that we observed descriptively (Figure 4.1) is repeated: Earnings rise until the late 40s and then decline for private-sector workers but remain constant for federal workers. The latter reflects the nature of the federal pay schedule, while the former is likely an artifact of worker behavior (e.g., older full-time workers might reduce their work time).

Figure 4.3
Pay Profiles Predicted Using the Current Population Survey and American Community Survey, 2003 Through 2012

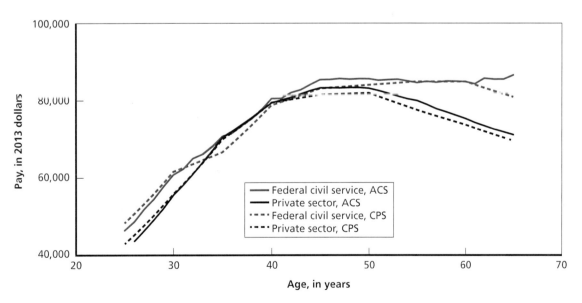

NOTE: The sample used for prediction in the figure includes men only with bachelor's degrees who work for wages or salaries, not self-employed, not veterans, and working full time (at least 35 hours per week) and full year (at least 40 weeks per year). Both CPS and ACS Tobit-adjusted results are weighted. See text earlier in this chapter for the models used to predict pay profiles.
RAND RR1503-4.3

We extend the model described above by allowing the estimated profile to take on different values for the return to education, age, and veteran status based on birth cohort. As mentioned, we pool the CPS for 1966 to 2014 and create synthetic birth cohorts. Appendix A includes additional details and regression results. Figures 4.4 and 4.5 present the predicted pay profiles using the model for a selected number of birth cohorts for the federal civil service and the private sector, respectively. Additionally, the figures include the ACS pay profile from Figure 4.3.

The private-sector pay profiles look remarkably similar to the pay profile predicted using cross-sectional data from the ACS. There is no distinctive trend in how the pay profiles change between birth cohorts that suggests a shift in the ages at which earning growth occurs. As described in greater detail in Appendix A, the results suggest a trend of declining returns from professional degrees among younger birth cohorts in the private sector. Also, veteran status, which generally has a negative association with pay, might have a zero or slightly positive association with pay among birth cohorts eligible for the draft (born before 1951).

Federal civil service pay profiles exhibit greater variation across birth cohorts than the private-sector pay profiles. Many of the birth cohorts in Figure 4.4 exhibit faster pay growth and greater overall earnings from ages 25 to 40 than those in Figure 4.3. However, changes in pay growth exhibit no discernable trend across birth cohorts. For example, between ages 25 and 30, pay growth is greater for the 1943 and 1983 birth cohorts than the ACS cross-sectional prediction, but the 1973 birth cohort exhibits lower earning growth. Perhaps the oddest pay profile in Figure 4.4 is the 1943 birth cohort, which has a substantial increase in federal pay between ages 35 and 40. Although this pay profile is not significantly different from the 1963 pay profile (see Table A.2 in Appendix A), pay remains persistently higher for the 1943 birth

Figure 4.4
Median Birth Cohort Pay Profiles for Federal Civil Service Workers

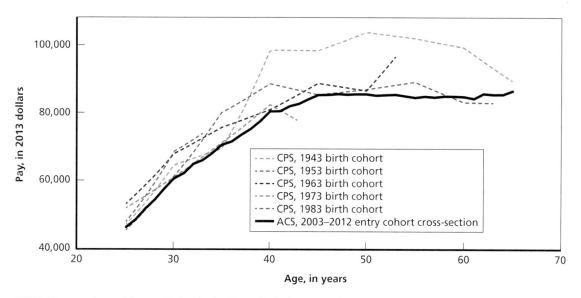

NOTE: The sample used for prediction in the figure includes men only with bachelor's degrees who work for wages or salaries, not self-employed, not veterans, and working full time (at least 35 hours per week) and full year (at least 40 weeks per year). Both CPS and ACS Tobit-adjusted results are weighted. See text earlier in this chapter for how we formed birth cohorts and the models used to predict pay profiles.
RAND RR1503-4.4

Figure 4.5
Median Birth Cohort Pay Profiles for Private-Sector Workers

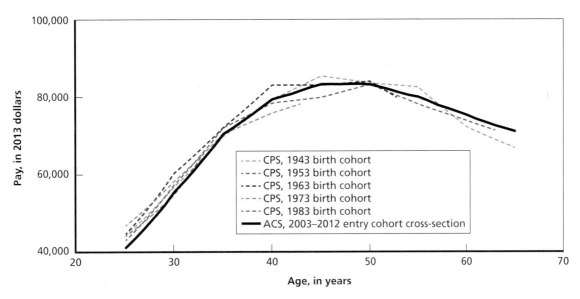

NOTE: The sample used for prediction in the figure includes men only with bachelor's degrees who work for wages or salaries, not self-employed, not veterans, and working full time (at least 35 hours per week) and full year (at least 40 weeks per year). Both CPS and ACS Tobit-adjusted results are weighted. See text earlier in this chapter for how we formed birth cohorts and the models used to predict pay profiles.
RAND *RR1503-4.5*

cohort after 40 (as well as for the 1948 cohort). This suggests that something might be characteristically different about these federal workers that the model does not capture. For example, it could be that these workers were able to retain greater pay rates during the period of high inflation in the early 1980s than new entrants into the federal civil service could.[3] Over time, salary adjustments might have affected specific parts of the pay schedule that our model could not capture by year effects alone. Sporadic adjustments to certain portions of the federal pay schedule might explain the birth cohort difference in the civil service pay profiles in Figure 4.4.

Discussion

The pay difference between the federal civil service and the private sector represents part of the worker's opportunity cost of staying in the civil service. Accurately representing the opportunity cost of staying is an important component of the DRM. If the opportunity cost of staying at younger or older ages were incorrectly specified in the model, that specification error would affect the parameter estimates and the simulations based on those estimates.

Here, we have been concerned with two issues. First, do pay profiles estimated from different samples have similar pay trajectories over a person's working life? After controlling for differences in the top-coding used in the CPS and in the ACS, we find that the predicted pay profiles from the two samples are similar for nonveteran men with only bachelor's degrees.

---

[3] The U.S. president controls federal civil service pay and determines changes to the pay schedule in conjunction with federal budgets that Congress approves.

Second, do pay profiles differ by birth cohort? Based on the raw data from the CPS, Figure 4.2 shows faster pay growth in the 1973 birth cohort than in the 1953 birth cohort. The difference was common to both federal and private-sector workers. After accounting for observed factors that are allowed to differ by birth cohort, we find that pay profiles for the private sector are very similar to the pay profiles derived using only a cross-section from the ACS. However, civil service pay profiles vary by birth cohort. We believe that differential adjustments to the federal pay schedule over time might cause these differences.

Our analysis of the pay of private-sector workers suggests that there is not a substantive difference between birth cohorts or between the pay profiles based on the CPS and the ACS. For the purposes of estimating the DRM in the next chapter, we use the cross-sectional pay profile for the private sector from the ACS because this is based on a larger sample of individuals. We alter the private-sector pay profile by not allowing pay to decline after age 46, the maximum projected earning level between the ages of 25 and 65. We do this assuming that the decline in pay is the result of workers choosing to work fewer hours, not employers reducing the pay of older workers.

Pay profiles for civil service workers are similar between the ACS and CPS after controlling for observable factors. Although life-cycle pay profiles differ by birth cohort, the differences are often not statistically significant. Still, the differences persist over repeated cross-sections, which suggests that they are real. Also, because of the smallness of the samples, the birth cohort pay profiles exhibit volatility over the life cycle. Because the DRM considers a forward-looking individual, the pay profile should reflect an individual's reasonable expectations about their future pay. We find it unlikely that individuals from the 1953 birth cohort would expect earnings to decrease in their early 40s or for people from the 1943 birth cohort to expect a sharp increase in pay between ages 35 and 40 (Figure 4.4). Further research will be needed to understand why these differences exist. We believe that a forward-looking individual would typically expect a more consistent life-cycle pay profile. As a result, we use the cross-sectional pay profile for the civil service from the ACS. Similar to what we did for the private sector, we alter the profile by not allowing pay to decline after age 46, the maximum projected earning level between the ages of 25 and 65.[4] The slight decline in pay between ages 47 and 60 might be the result of some decrease in hours of work supplied.

Figure 4.6 presents the final private-sector and civil service pay profiles. The pay profiles are close together. Civil service pay exceeds private-sector pay by about $5,000 per year for ages 25 to 35. The pays are nearly equal for ages 35 to 45, and, beyond age 45, civil service pay is $2,350, or 2.8 percent, higher.

---

[4]   We exclude the slight increase in the pay profile from this calculation that occurs in the 60s.

Figure 4.6
U.S. Department of Defense Civil Service and Private-Sector Pay Profiles Used in the Dynamic Retention Model

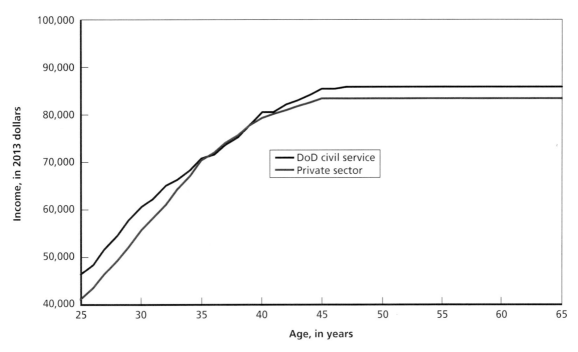

CHAPTER FIVE

Dynamic Retention Model Estimation Results

We estimate DRMs for each entry cohort (1988 through 2000) and one that combines the data across cohorts. This analysis allows us to test whether parameter estimates differ by entry cohort and, if so, whether the predicted responses to changes in compensation policies are substantively different. We also extend the DRM to allow the taste distribution to differ between veterans and nonveterans. The first part of the chapter presents the entry cohort–specific and combined model results. The second part explores evidence of changing tastes for civilian service across entry cohorts and the implications for retention responsiveness to compensation policy.

Dynamic Retention Model Estimates, by Entry Cohort

Tables 5.1 and 5.2 show the parameter estimates and standard errors (SEs) of the DRM for each entry cohort. Because we estimate the discount factor, the probability of censoring and of dropping out in the first year as a logit, we transform the estimates into probabilities after estimation. Table 5.3 shows the transformed parameter estimates.[1] Figure 5.1 addresses model fits. We first discuss the model fits and then specific parameter estimates.

Model Fits

For each entry cohort, Figure 5.1 compares the observed cumulative retention rate by year of service in the data (black line) and the predicted cumulative retention rate that we simulated using the estimated model parameters (red line) and assuming the current baseline compensation system. The dotted lines are error bands associated with the cumulative retention rates in the data, computed using the Kaplan–Meier estimator. Because our data end in 2013, we have more years over which to track earlier entry cohorts than later ones. Thus, Figure 5.1 compares the predicted cumulative retention curve and the actual data over 25 years for the 1988 entry cohort and over 13 years for the 2000 cohort.

The models fit well, and the parameter estimates are all statistically significantly different from 0. Comparison of the simulated and actual retention profiles show that the simulated retention profiles generally fall within the error bands. Thus, the retention profiles that the models predict are close to the actual data.

---

[1] Some of the parameters are transformed using a logit function if they are to be bounded between 0 and 1. The transformations are important during estimation because our optimization function does not account for these constraints. We calculate SEs only on the untransformed parameter because that is where they are meaningful. However, during our exposition, we focus on the transformed parameters because these are economically meaningful.

Table 5.1
Parameter Estimates and Standard Errors, by Entry Cohort Entry Year, 1988 Through 1994

Parameter	1988		1989		1990		1991		1992		1993		1994	
	Estimate	SE	Estimate	SE	Estimate	SE	Estimate	SE	Estimate	SE	Estimate	SE	Estimate	SE
Shock scale, λ	57.34	2.47	58.67	1.94	56.65	2.95	51.57	2.55	46.10	2.52	48.00	4.17	43.06	3.37
Mean of nonveterans' taste distribution, μ_{Nonvet}	10.28	2.58	6.21	0.52	−0.85	1.88	−7.17	0.37	−8.88	0.37	−6.60	2.39	−9.31	1.55
Mean of veterans' taste distribution, μ_{Vet}	1.42	1.15	0.85	0.76	0.30	0.77	−0.33	0.96	1.28	0.77	5.13	1.25	3.98	0.83
SD of nonveterans' taste distribution, σ_{Nonvet}	21.40	2.08	15.74	0.76	12.50	1.20	10.39	0.97	11.06	1.07	16.70	2.07	15.09	1.67
SD of veterans' taste distribution, σ_{Vet}	−10.81	2.35	−6.13	1.57	−16.42	2.88	−0.10	1.76	−2.03	1.59	−3.54	2.19	−4.50	1.65
Accounting for the censoring until 1995, θ	−2.63	0.08	−3.29	0.10	−3.20	0.13	−2.47	0.09	−3.92	0.33	−2.82	0.19	−3.34	0.25
Shift in the probability of staying in the first year due to early attrition, δ	−2.69	0.10	−2.71	0.07	−3.10	0.13	−3.33	0.19	−3.27	0.21	−3.44	0.35	−3.66	0.37
Personal discount factor, β	2.16	0.11	2.37	0.01	2.63	0.08	2.79	0.01	2.79	0.01	2.59	0.12	2.71	0.08
Log likelihood	−18,148		−31,222		−14,828		−13,614		−10,564		−6,885		−7,410	
Sample size	7,363		12,758		6,167		5,638		4,495		2,976		3,296	

NOTE: The mean of the taste distribution of veterans is given by $\mu_{Vet} + \mu_{Nonvet}$, and the SD of the veterans' taste distribution is given by $\sigma_{Vet} + \sigma_{Nonvet}$. The SEs can be used to test the hypothesis that mean taste of veterans is statistically significantly different from the mean of the taste distribution for nonveterans. For example, in the case of 1988, with a mean taste of 1.42 (making the total mean of the veterans' taste distribution 1.42 + 10.28) and SE of 1.15, we would say that mean taste of veterans is not statistically significantly different from that of nonveterans at the 5-percent level. However, in 1994, mean taste of veterans is 3.98 (making the total mean of the veterans' taste distribution 3.98 + 15.09), and the SE is 0.83. This indicates that we can reject the null hypothesis that the mean of the veterans' taste distribution is the same as the mean of nonveterans' taste distribution at the 5-percent level. This process also holds for the parameter estimates of the SD. In 1988, the SD of the veterans' taste distribution is −10.81 (making the total SD −10.81 + 21.40) with an SE of 2.35, indicating that it is significantly less than the SD for nonveterans.

Table 5.2
Parameter Estimates and Standard Errors, by Entry Cohort Entry Year, 1995 Through 2000

Parameter	1995 Estimate	1995 SE	1996 Estimate	1996 SE	1997 Estimate	1997 SE	1998 Estimate	1998 SE	1999 Estimate	1999 SE	2000 Estimate	2000 SE
Shock scale, λ	43.40	3.74	42.30	3.72	38.35	3.95	45.07	5.78	69.65	13.77	49.91	7.15
Mean of nonveterans' taste distribution, μ	−5.61	1.59	−0.04	1.87	1.46	2.28	−2.27	2.45	0.19	2.62	−0.46	2.60
Mean of veterans' taste distribution, μ	3.70	0.93	1.62	0.96	2.93	1.17	2.04	1.05	0.10	1.47	2.88	1.14
SD of nonveterans' taste distribution, σ	18.44	2.16	19.32	2.31	18.61	2.83	17.28	3.01	25.14	6.27	17.76	3.15
SD of veterans' taste distribution, σ	−5.30	1.88	−2.86	1.96	−2.51	2.33	−6.27	2.64	−11.19	5.25	−6.66	3.07
Accounting for the censoring until 1995, θ	−4.22	0.71	0.00	0.00	0.00	0.00	0.00	0.00	0.00	0.00	0.00	0.00
Shift in the probability of staying in the first year due to early attrition, δ	−4.17	0.76	−4.30	0.63	−3.25	0.31	−3.07	0.27	−2.73	0.18	−3.05	0.18
Personal discount factor, β	2.54	0.09	2.39	0.09	2.31	0.12	2.55	0.12	2.79	0.02	2.66	0.11
Log likelihood	−7,577		−6,681		−5,488		−4,545		−4,824		−6,379	
Sample size	3,513		3,249		2,808		2,327		2,729		3,810	

NOTE: The mean of the taste distribution of veterans is given by $\mu_{Vet.} + \mu_{Nonvet.}$, and the SD of the veterans' taste distribution is given by $\sigma_{Vet.} + \sigma_{Nonvet.}$. The SEs can be used to test the hypothesis that mean taste of veterans is statistically significantly different from the mean of the taste distribution for nonveterans. For example, in the case of 1988, with a mean taste of 1.42 (making the total mean of the veterans' taste distribution 1.42 + 10.28) and SE of 1.15, we would say that mean taste of veterans is not statistically significantly different from that of nonveterans at the 5-percent level. However, in 1994, mean taste of veterans is 3.98 (making the total mean of the veterans' taste distribution 3.98 + 15.09), and the SE is 0.83. This indicates that we can reject the null hypothesis that the mean of the veterans' taste distribution is the same as the mean of nonveterans' taste distribution at the 5-percent level. This process also holds for the parameter estimates of the SD. In 1988, the SD of the veterans' taste distribution is −10.81 (making the total SD −10.81 + 21.40) with an SE of 2.35, indicating that it is significantly less than the SD for nonveterans.

Table 5.3
Transformed Parameter Estimates and Policy Response, by Entry Cohort Entry Year

Parameter	1988	1989	1990	1991	1992	1993	1994	1995	1996	1997	1998	1999	2000
Shock scale, λ	57.34	58.67	56.65	51.57	46.10	48.00	43.06	43.40	42.30	38.35	45.07	69.65	49.91
Mean of nonveterans' taste distribution, μ	10.28	6.21	−0.85	−7.17	−8.88	−6.60	−9.31	−5.61	−0.04	1.46	−2.27	0.19	−0.46
Mean of veterans' taste distribution, μ	1.42	0.85	0.30	−0.33	1.28	5.13	3.98	3.70	1.62	2.93	2.04	0.10	2.88
SD of nonveterans' taste distribution, σ	21.40	15.74	12.50	10.39	11.06	16.70	15.09	18.44	19.32	18.61	17.28	25.14	17.76
SD of veterans' taste distribution, σ	−10.81	−6.13	−16.42	−0.10	−2.03	−3.54	−4.50	−5.30	−2.86	−2.51	−6.27	−11.19	−6.66
Accounting for the censoring until 1995, θ	0.07	0.04	0.04	0.08	0.02	0.06	0.03	0.01	0.00	0.00	0.00	0.00	0.00
Shift in the probability of staying in the first year due to early attrition, δ	0.06	0.06	0.04	0.03	0.04	0.03	0.03	0.02	0.01	0.04	0.04	0.06	0.05
Personal discount factor, β	0.90	0.91	0.93	0.94	0.94	0.93	0.94	0.93	0.92	0.91	0.93	0.94	0.93
Policy response													
Percentage change in force size as a result of a 1-percent across-the-board decrease in real pay	−2.1	−2.9	−3.8	−5.8	−5.3	−3.6	−4.5	−3.7	−3.0	−3.0	−3.7	−2.2	−3.2

NOTE: We transformed the parameters β, δ, and θ using a logit specification, which bounds the outcome between 0 and 1. Some of the parameters are transformed using a logit function if they are to be bounded between 0 and 1. The transformations are important during estimation because our optimization function does not account for these constraints. We calculate SEs only on the untransformed parameter because that is where they are meaningful. However, during our exposition, we focus on the transformed parameters because these are economically meaningful.

Figure 5.1
Model Fits, U.S. Department of Defense Civil Service Retention, 1988–2000 Entry Cohorts

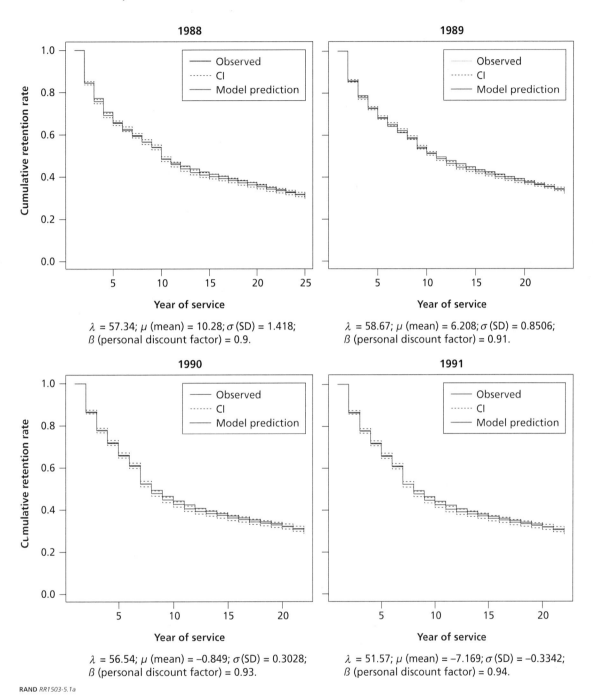

λ = 57.34; μ (mean) = 10.28; σ (SD) = 1.418;
β (personal discount factor) = 0.9.

λ = 58.67; μ (mean) = 6.208; σ (SD) = 0.8506;
β (personal discount factor) = 0.91.

λ = 56.54; μ (mean) = –0.849; σ (SD) = 0.3028;
β (personal discount factor) = 0.93.

λ = 51.57; μ (mean) = –7.169; σ (SD) = –0.3342;
β (personal discount factor) = 0.94.

The panels for 1988 through 1996 in Figure 5.1 also illustrate the censoring problem at year 9 in 1988, year 8 in 1989, year 7 in 1990, and so forth, when retention falls by an expectedly large amount, shown by a kink in the retention profile at that year. As discussed in Asch, Mattock, and Hosek, 2014a, the drop corresponds to the disappearance of personnel in some occupations, especially cartography, from the DoD civil service in 1997. This occurred because of a reorganization that caused the export of some DoD functions to a newly created agency

Figure 5.1—Continued

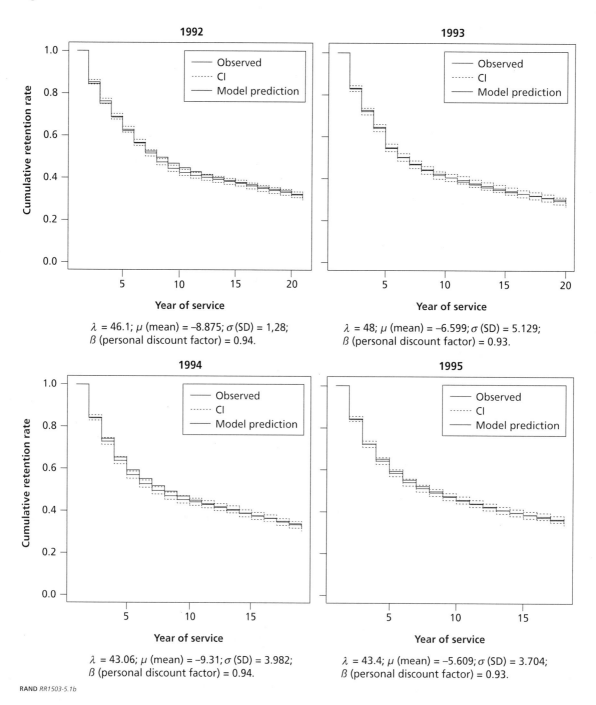

λ = 46.1; μ (mean) = –8.875; σ (SD) = 1,28;
β (personal discount factor) = 0.94.

λ = 48; μ (mean) = –6.599; σ (SD) = 5.129;
β (personal discount factor) = 0.93.

λ = 43.06; μ (mean) = –9.31; σ (SD) = 3.982;
β (personal discount factor) = 0.94.

λ = 43.4; μ (mean) = –5.609; σ (SD) = 3.704;
β (personal discount factor) = 0.93.

RAND *RR1503-5.1b*

external to DoD. Specifically, in 1997, the National Imagery and Mapping Agency, a major employer of cartographers, was created by combining various defense functions, including the Defense Mapping Agency. Adding the parameter θ for the 1988–1995 entry cohorts allowed us to retain the censored observations and account for the censoring at this point when we estimated the models. The model fits are excellent, regardless of whether we include or exclude the additional parameters and the censored observations. However, as shown in Tables 5.1 and 5.2, the estimated censoring parameter (θ) is positive and statistically significantly different

Figure 5.1—Continued

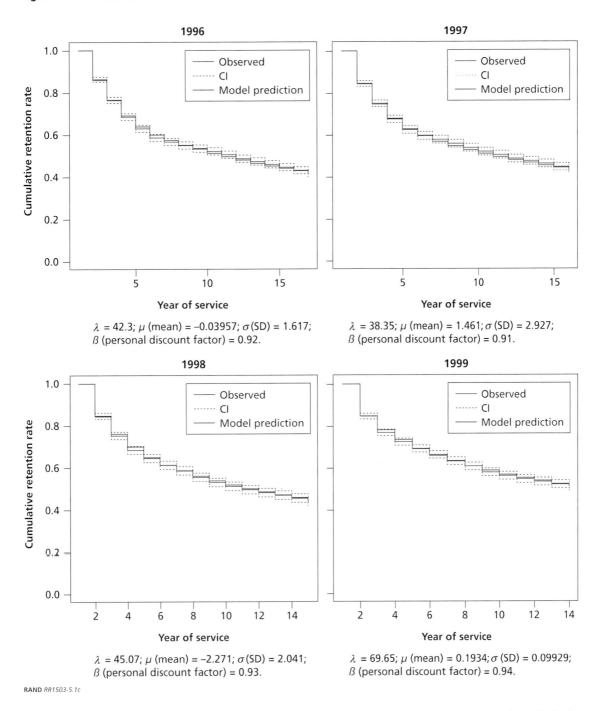

1996

λ = 42.3; μ (mean) = –0.03957; σ(SD) = 1.617; β (personal discount factor) = 0.92.

1997

λ = 38.35; μ (mean) = 1.461; σ (SD) = 2.927; β (personal discount factor) = 0.91.

1998

λ = 45.07; μ (mean) = –2.271; σ(SD) = 2.041; β (personal discount factor) = 0.93.

1999

λ = 69.65; μ (mean) = 0.1934; σ (SD) = 0.09929; β (personal discount factor) = 0.94.

RAND RR1503-5.1c

from 0 in the models for 1988 through 1995. For example, censoring is estimated to shift the retention profile by between 6.3 percent for the GS employees with at least bachelor's degrees in the 1988 entry cohort and by 1.5 percent for those entering in 1995.

Parameter Estimates for Entry Cohort–Specific Models

Tables 5.1 and 5.2 present the parameter estimates and SE for each entry cohort, while Table 5.3 shows the transformed parameter estimates. The estimated scale parameter, λ, and

Figure 5.1—Continued

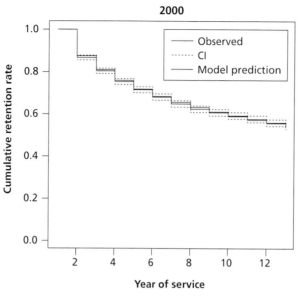

2000

λ = 49.91; μ (mean) = –0.4462; σ (SD) = 2.883;
β (personal discount factor) = 0.93.

NOTE: CI = confidence interval.

RAND *RR1503-5.1d*

the estimated mean and SD of the taste distribution, μ and σ, respectively, are denominated in thousands of dollars. Thus, the estimate of 10.28 for μ for nonveterans in the 1988 entry cohort in Tables 5.1 through 5.3 represents $10,280.

The estimated SDs of the shock term are relatively large. Because the variance of the shock term is related to λ according to the formula

$$\frac{\pi^2 \lambda^2}{6},$$

the SD of the shock is

$$\frac{\pi \lambda}{\sqrt{6}}.$$

The estimated SDs are, for instance, $75,660 for the 1989 entry cohort, $49,180 for the 1997 entry cohort, and $64,016 for the 2000 entry cohort. The large estimate for the 1999 entry cohort, $89,335, appears to be an outlier.

Like we did in Asch, Mattock, and Hosek, 2014a, we find that estimated mean taste is positive for the 1988 entry cohort. In the earlier study, estimated mean taste was $13,970. In our updated model, we estimate a mean taste of $10,280.[2] The samples differ slightly, with the

---

[2]   The $10,280 figure is for nonveterans. However, in the 1988 cohort, virtually all entrants are nonveterans, so the $10,280 is generally representative of mean taste for the entire entry cohort.

sample in this report including people with professional degrees. A positive mean taste implies that civil service workers, on average, place a positive value on civil service employment over and above the monetary aspects included in the model. For example, they might value the non-monetary aspects of employment that offers an opportunity to serve the public or job security. In addition, they might value nonwage benefits of the job, such as health insurance, that might be better in the civil service than in the private sector.

Also like in our earlier study, we find heterogeneous tastes for DoD civil service for the 1988 entry cohort. Our updated estimate of the SD of taste for the 1988 entry cohort is virtually identical to our earlier estimate, $21,400 versus $21,940, in our earlier study. Like in the earlier analysis, the estimated SD exceeds the estimated mean of the taste distribution. A relatively large SD means that, controlling for compensation and shocks, there will be considerable variation in the stay/leave decisions of civil service workers at a given point in their careers.

A comparison of the estimated mean tastes across entry cohorts in Tables 5.1 through 5.3 shows that the positive estimate for the 1988 cohort is an exception, at least for nonveterans, rather than the rule among the cohorts we consider. To illustrate the differences, Figure 5.2 shows a plot of the estimated means and SDs of the taste distribution for nonveterans and veterans, across entry cohorts. Estimated mean taste for nonveterans (solid blue line) declines from $10,280 for the 1988 entry cohort to $6,208 for the 1989 cohort (statistically significantly different from 0) and then declines further to $850 for the 1990 entry cohort (not statistically significantly different from 0). Estimated mean tastes for nonveterans then become negative, to –$7,170 for the 1991 entry cohort and –$8,875 for the 1992 cohort. Mean taste remains at around –$8,000 between 1992 and 1994 but then begins to increase, up to –$5,610 in 1995 and then to $0 by 1996. For the 1996–2000 entry cohorts, estimated mean taste for nonveterans is not statistically significantly different from 0. In short, estimated mean taste decreases between the 1988 and 1991 entry cohorts, levels out at a negative level until 1994, and then increases after that, but not back up to the 1988 level. Instead, mean taste is around $0 after 1995 for nonveterans.

Given the generally negative or 0 estimated mean tastes after 1989, it is clear that the positive and relatively high estimates for 1988 and 1989 are outliers. One possible explanation for this is the change in the overall size of the civilian workforce in DoD. Civilian employment in DoD increased during the 1980s, reaching a peak of 1.1 million in 1985 and staying at that high level until 1989 (OPM, undated [a]). DoD employment then fell precipitously with the end of the Cold War and the resulting drawdown. Between 1989 and 1990, employment fell by 40,000 workers in just one year. Civilian employment continued to fall rapidly during the early 1990s and then fell more slowly in the mid-1990s, reaching about 650,000 in 2001, and staying at about that level through 2007. Employment has increased somewhat since then, up to 723,000 in 2014, the last year for which there are data. To speculate, the high and positive mean taste for the late 1980s might reflect a Cold War, high-employment regime, and the negative taste from 1991 to 1995 might result in part from reductions in force caused by downsizing in the early 1990s.

The estimated SD of tastes for nonveterans (solid green line) also varies across entry cohorts, although the time pattern differs somewhat from the pattern for estimated mean tastes. The estimated SD declines between the 1988 and 1992 entry cohorts, but, rather than stay relatively level between 1992 and 1994, it begins to rise again, reaching the 1988 level of around $21,000 for entry cohorts after 1994 and staying roughly at that level through 2000.

Figure 5.2
Estimated Parameters of the Taste Distribution, by Entry Cohort

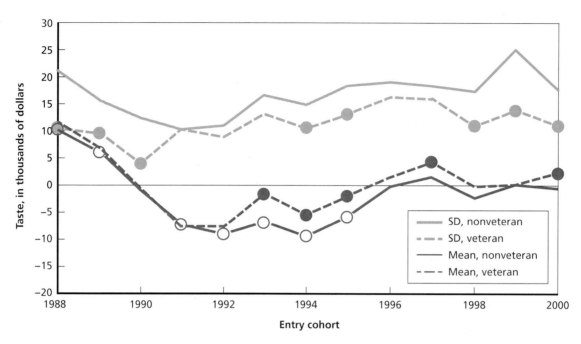

NOTE: Parameter estimates reflect those presented in Tables 5.1 and 5.2. If veterans' mean taste is statistically significantly different from 0 at the 5-percent level, a hollow circle denotes it. All SDs for nonveterans are statistically significantly different from 0. If the mean and SD of veterans' taste distribution are statistically significantly different from those of nonveterans, we denote it with a solid circle.
RAND *RR1503-5.2*

We find a difference in the taste distributions of veterans and nonveterans. Consistently with our hypothesis, we find that, for most years, veterans have a higher mean taste for DoD civil service than nonveterans and a lower taste variance. As shown in Figure 2.3 in Chapter Two, veterans became a sizable portion of entrants only after 1991. We find that veterans' mean taste is not statistically significantly different from nonveterans' mean tastes until the 1993 entry cohort. As seen in Figure 5.2, veterans' mean tastes are then higher. The difference varies across entry cohorts, but veterans' mean tastes are about $3,000 higher than those of nonveterans. On the other hand, we find that mean tastes among veterans are less heterogeneous. Although the difference between the estimated SDs for veterans and nonveterans varies across entry cohorts, the estimate is about $5,400 less after 1992. These findings accord with the view that veterans, many of whom are military retirees, have already revealed a taste for DoD employment over external opportunities, as evidenced by their experience as members of the armed forces.

The estimated personal discount factor, β, ranges from 0.90 to 0.94 across entry cohorts, with the higher rates for the 1991–1992 cohorts and the 1999–2000 entry cohorts. A discount factor of 0.90 means that $1.00 received in one year is worth $0.90 today. The estimate of 0.90 for the 1988 entry cohort is consistent with our earlier study, in which we also estimated a 0.90 discount factor.

The estimated model also includes a parameter to capture the shift in the probability of staying in the first year because of early attrition, δ. The fit graphics (Figure 5.1) show the first-year drop in retention, and the estimated δ is positive and statistically significantly different from 0 in every model. The estimate is largest for the 1988 entry cohort (0.063) and smallest for the 1996 entry cohort (0.013), with average around 0.03 or about 3 percent.

Figure 5.3 illustrates an interesting regularity in our estimates—namely, the estimated mean taste and discount factor are inversely related, with a correlation coefficient of –0.79. Mean taste is on the left axis, while the personal discount rate is measured on the right. Other factors that are correlated with both the personal discount factor and mean taste, such as age and education, might cause this negative correlation, and controlling for them could eliminate this correlation. One approach to addressing this correlation is to add covariates that can shift mean taste (like we do with veteran status). Another approach is to set the discount factor at a fixed value that does not vary with entry cohort year. Because of the additional computational burden of adding covariates, we adopt the approach of fixing the personal discount factor. In particular, we reestimated the models in Tables 5.1 and 5.2 with a fixed value of the discount factor such that the value we chose was the estimated personal discount factor in the DRM that combines data for entry cohorts 1993 to 2000. As we show in Table 5.4, the estimated personal discount factor in the combined model is 0.93. The parameter estimates for the models with fixed personal discount factor are reported in Tables B.1 through B.3 in Appendix B. As we show there, however, our results are qualitatively similar to those shown in Tables 5.1 through 5.3 and Figure 5.1, in which the personal discount factor is not fixed. The notable exception is that we no longer find positive and relatively large mean taste in 1988 and 1989, lending further support that the estimated mean tastes for these Cold War entry cohorts are more anomalous than the estimated mean tastes for later entry cohorts, although the source of the anomaly is unclear.

Figure 5.3
Estimates of Nonveterans' Mean Taste and Personal Discount Factor, by Entry Cohort

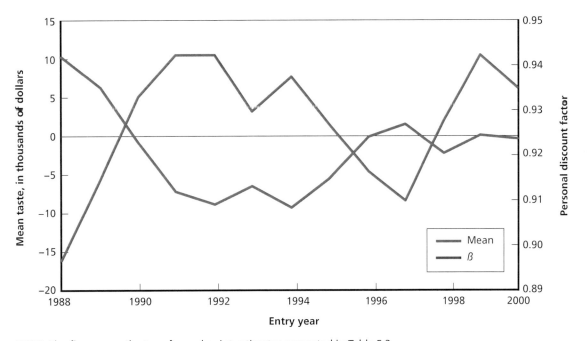

NOTE: The figure uses the transformed point estimates presented in Table 5.3.
RAND RR1503-5.3

Implications for Responsiveness to Pay Changes

The estimates in Tables 5.1 through 5.3 and in Tables B.1 through B.3 in Appendix B (in which we fix the personal discount factor) together with the patterns observed in Figures 5.2 and 5.3 suggest that our estimates vary to some degree across entry cohorts. Two questions emerge from this observation. First, do these results imply that the more-recent entrants have different tastes for civil service from those of earlier entrants? Second, does the responsiveness to changes in compensation differ across entry cohorts? We explore the answer to the first question after we discuss the DRM results from a model that combines data from all entry cohorts, which we use as a benchmark for comparing the entry cohort–specific results. In this subsection, we focus on the second question: what our results imply for how the responsiveness to changes in compensation differs across cohorts.

Using the estimates for each cohort, we simulated the steady-state civil service retention profile assuming a 1-percent real, across-the-board decrease in civil service pay and computed the percentage change in retention across the force. The last row of Table 5.3 shows the results when we use models in which the personal discount rate is estimated. Similarly, the last row of Table B.3 in Appendix B shows results when we use models in which it is fixed at 0.93. In Figure 5.4, we graph the percentage change in force size from these two types of models.

Figure 5.4 shows that the responsiveness to a pay change was greatest for the 1991 and 1992 entry cohorts. That is, the decrease in the force retained was larger for those cohorts than either for the earlier Cold War entry cohorts of 1988 and 1989 and the postdrawdown cohorts after 1995. For example, we simulate that a 1-percent decrease in pay reduces the force by 2.9 percent for the 1989 cohort but by 5.8 percent for the 1991 cohort. It is possible that,

Figure 5.4
Simulated Percentage Change in Steady-State Force Size

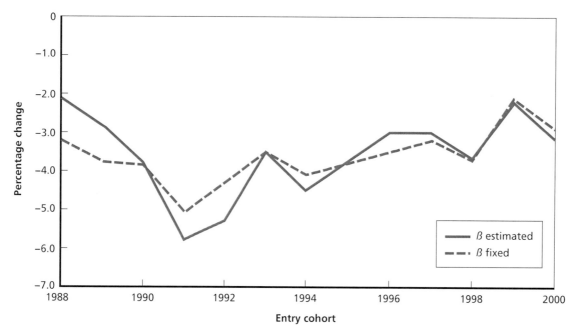

NOTE: This figure shows the simulated percentage change in steady-state force size due to a 1-percent real decrease in civil service pay using DRMs in which the personal discount factor, β is estimated or, alternatively, fixed.
RAND RR1503-5.4

during the years when the drawdown was particularly steep, civilians entering during these years felt less secure about their employment, so we find greater responsiveness for these entry cohorts. That said, the simulated drop in force size as a result of a 1-percent decrease in pay is relatively stable at around 3.5 percent after 1995 in the models in which we estimate the personal discount factor and after 1992 in the models in which the personal discount factor is fixed. Furthermore, even with the variability shown in Figure 5.4, retention responsiveness to pay is broadly similar across entry cohorts, ranging from –2.2 percent to –5.1 percent in the model in which the personal discount factor is fixed. Although this difference is not trivial, it is also not orders of magnitude either. We also see that the responsiveness to a pay change is about the same for entry cohorts entering after 1992 regardless of whether the personal discount factor is fixed or estimated.

The percentage change in the steady-state force size in response to a change in pay is related to the estimated parameters of the taste distribution. Responsiveness is correlated with both the estimated mean and SD of the taste distribution. The correlation coefficients between force-size change and the mean and SD of taste are 0.79 and 0.87, respectively, in the model in which β is estimated. In the models in which we fix the personal discount factor at 0.93, the correlation coefficients are 0.56 and 0.86, respectively. To illustrate the relationship with the taste distribution, Figure 5.5 shows the estimated SD of the taste distribution and the simulated percentage change in steady-state force size from a 1-percent drop in pay, using models in which we fix the personal discount factor (shown in Appendix B).

The strong correlation between pay responsiveness and taste SD implies that, when tastes are more homogeneous, pay responsiveness is greater. Intuitively, when civilian employees have

Figure 5.5
Estimated Standard Deviation of Veterans' Taste and Percentage Change in Steady-State Force Size, by Entry Cohort

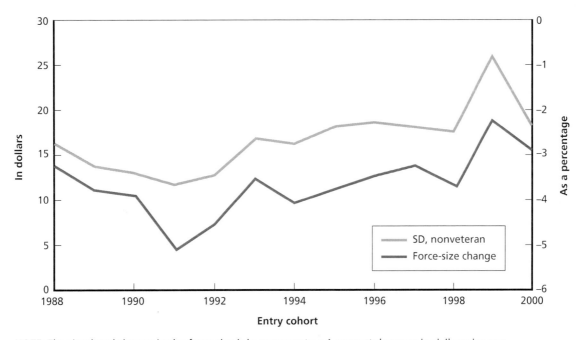

NOTE: The simulated change in the force size is in response to a 1-percent decrease in civil service pay.
RAND RR1503-5.5

similar tastes for public service in DoD, they are more likely to respond in a similar fashion when pay changes, so changes in retention are larger than when civilian employees have more-diverse tastes for service. Also, in a manner that is consistent with intuition, when mean tastes are higher, pay responsiveness is reduced. We illustrate this correlation with mean taste in Figure 5.6.

Dynamic Retention Model Estimates Combining Entry Cohorts in a Single Model

Estimating separate models for each entry cohort, like we show in Tables 5.1 through 5.3 (and in Tables B.1 through B.3 in Appendix B, in which we fix the personal discount factor), allows us to examine how our estimates for each DRM parameter change across entry cohorts. However, for analysis of the retention and cost effects of compensation changes, it can be unruly to show results from multiple models rather than a single model. Variations in parameter estimates across entry cohorts suggest that singling out any cohort might not be the most sensible path. Instead, we estimate a single model in which we combine data from all of the entry cohorts. This has the advantage of incorporating information from each entry cohort while having the simplicity of a single model.

The single model is specified to allow the mean and SD of taste to differ between pre-1992 and post-1991 entry cohorts and by veteran status. As discussed in the context of Figure 5.2, mean tastes decline between 1988 and 1992. Thus, we have estimated the mean and SD of tastes for nonveterans and veterans for pre-1992 entry cohorts and for post-1991 entry cohorts.

Figure 5.6
Estimated Mean of Nonveterans' Taste and Percentage Change in Steady-State Force Size, by Entry Cohort

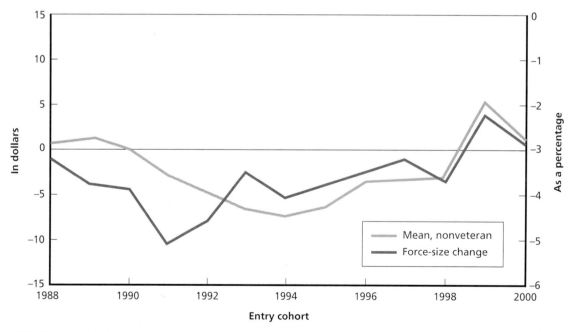

NOTE: The simulated change in the force size is in response to a 1-percent decrease in civil service pay.
RAND *RR1503-5.6*

We estimate a single shock parameter, λ, a single personal discount rate, a single parameter to account for pre-1996 censoring, and a single parameter to account for first-year attrition.

Table 5.4 shows the results. The first three columns show results in which we combine the 1988–2000 entry cohorts, while the last three columns show results in which we exclude the pre-1992 entry cohorts and combine the 1992–2000 entry cohorts. Figures 5.7 and 5.8 show the model fits for each of these models. As can be seen, the fit is quite good in both cases but better with the model that excludes pre-1992 cohorts and estimates with data from the 1992–2000 entry cohorts only. When we conduct policy simulations in Chapter Six, we rely on the

Table 5.4
Parameter Estimates and Standard Errors, Models That Combine Entry Cohorts

Parameter	Entry Cohorts 1988–2000			Entry Cohorts 1992–2000		
	Estimate	SE	Transformed Estimate	Estimate	SE	Transformed Estimate
Location, λ	53.78	0.93	53.78	45.15	1.36	45.15
Mean of nonveterans' taste distribution 1988–1991, μ	−0.95	0.57	−0.95	N/A	N/A	N/A
Mean of nonveterans' taste distribution 1992+, μ	−1.84	0.20	−1.84	−4.16	0.63	−4.16
Mean of veterans' taste distribution 1988–1991, μ	0.58	0.42	0.58	N/A	N/A	N/A
Mean of veterans' taste distribution 1992+, μ	3.57	0.34	3.57	2.95	0.32	2.95
SD of nonveterans' taste distribution 1988–1991, σ	12.54	0.42	12.54	N/A	N/A	N/A
SD of nonveterans' taste distribution 1992+, σ	6.84	0.43	6.84	16.54	0.69	16.54
SD of veterans' taste distribution 1988–1991, σ	−3.82	0.81	−3.82			
SD of veterans' taste distribution 1992+, σ	−6.38	0.72	−6.38	−4.13	0.69	−4.13
Accounting for the censoring until 1995, θ	0.01	0.00	0.01	0.03	0.00	0.03
Shift in the probability of staying in the first year due to early attrition, δ	−3.07	0.05	0.04	−3.42	0.11	0.03
Personal discount factor, β	2.58	0.03	0.93	2.59	0.03	0.93
Log likelihood	−138,973			−60,662		
Sample size	61,129			29,203		
Policy response						
Percentage change in steady-state force size due to a 1% drop in pay	−3.98			−3.74		

Figure 5.7
Model Fit, Single Model Combining 1988–2000
Entry Cohort Data: U.S. Department of Defense Civil
Service Retention

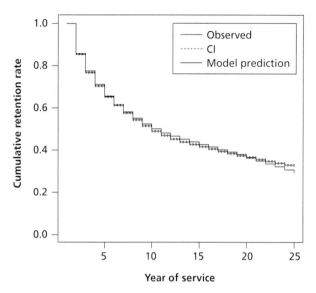

λ = 53.78; μ (mean) = –0.9497; σ (SD) = –1.842;
β (personal discount factor) = 0.93.

RAND *RR1503-5.7*

Figure 5.8
Model Fit, Single Model Combining 1992–2000
Entry Cohort Data: U.S. Department of Defense Civil
Service Retention

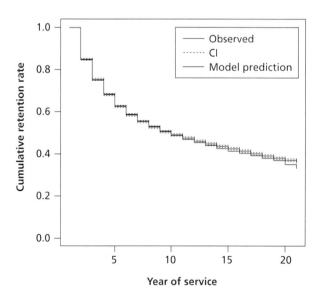

λ = 45.15; μ (mean) = –4.159; σ (SD) = 2.952;
β (personal discount factor) = 0.93.

RAND *RR1503-5.8*

single model that combines the 1992–2000 entry cohorts, shown in the final three columns of Table 5.4.

The results for the combined models are consistent with the entry cohort–specific results in Tables B.1 through B.3 in Appendix B, in which we fix the personal discount factor at 0.93, the value we estimated in the single model. We estimate a personal discount factor of 0.93 for both the 1988–2000 combined model and the 1992–2000 combined model. In the 1992–2000 combined model, we estimate a mean taste of –$4,160 and an SD of taste of $16,540. Like in the entry cohort–specific models, we find that veterans have higher mean tastes but less heterogeneous tastes (lower SD) and that these differences are statistically significant from 0.

Exploring Whether Tastes Vary Across Entry Cohorts

The parameter estimates by entry cohort can be used to address the question of whether these results imply that the more-recent entrants have different tastes for civil service from those of earlier entrants. As discussed in Chapter Three, estimated mean tastes might vary across entry cohorts because of changes in the mean taste of those entering or because of changes in an entry cohort's demographic composition. The question we consider in this subsection is whether the mean taste is different for those entering in later calendar years. In doing this, we estimate a model in which mean taste is allowed to change from entry cohort to entry cohort. Importantly, we control for demographic composition by using a subsample with homogeneous, narrowly defined observed characteristics. Also, we use a combined model specification in which we constrain the other parameters—namely, the SD of taste, personal discount factor, early attrition, and censoring parameters—to be the same across cohorts.

Specifically, we estimated a model that combines data across entry cohorts for a sample that consists of male nonveterans with only bachelor's degrees who are ages 30 or younger, entering between 1992 and 2000.[3]

Table 5.5 shows the parameter estimates and the transformed estimates. To explore whether mean taste varies across entry cohorts, the model we estimate allows mean taste to shift with entry cohort year. For example, we estimate mean taste for the 1992 entry cohort of –$14,994 and a change in mean taste in 1993 of –$2,508. Consequently, as seen in the final column, "Transformed Estimate," the estimated mean taste for the 1993 entry cohort is –$17,502 (–$14,994 – $2,508).

We find some evidence that mean taste differs across entry cohorts. That is, some differences across cohorts in mean tastes are statistically significant. The –$14,994 estimate of mean taste in 1992 is statistically significantly different from 0, as is the –$2,508 change in 1993. Thus, mean taste became more negative in 1993. However, the changes for the 1994 and 1995 entry cohorts are not statistically significant, implying that tastes were relatively stable at around –$17,000 between 1992 and 1995. We estimate mean tastes to have increased for the 1996 entry cohort, by $3,521 (a statistically significant difference from 0) up to –$14,301, a

---

[3] In this exploratory analysis, we do not adjust the civil service or external pay profiles to reflect the demographics of this subgroup (bachelor's degree only) but use the same pay profiles as we used for the overall sample. We estimate mean tastes to be uniformly lower than shown for the overall sample in Tables 5.3 and 5.4 because the taste parameter captures permanent differences in earning levels and the average level of bachelor's-degree earnings will be lower than the average level of bachelor's-degree-or-higher earnings.

Table 5.5
Parameter Estimates and Standard Errors, Combined Entry Cohorts 1992–2000, Male, Nonveteran Bachelor's Degree Holders Entering at Age 30 or Younger

Parameter	Estimate	SE	Transformed Estimate
Location, λ	53.83	6.69	53.83
Mean of the taste distribution, 1992	–14.99	2.65	–14.99
Change in mean of the taste distribution, 1993	–2.51	1.06	–17.50
Change in mean taste distribution, 1994	1.119	1.15	–16.38
Change in mean taste distribution, 1995	–1.33	1.14	–17.71
Change in mean taste distribution, 1996	3.52	1.21	–14.19
Change in mean taste distribution, 1997	1.31	1.14	–12.88
Change in mean taste distribution, 1998	–1.42	1.26	–14.30
Change in mean taste distribution, 1999	3.27	1.35	–11.03
Change in mean taste distribution, 2000	0.36	1.11	–10.67
SD of the taste distribution for nonveterans, σ	15.07	3.01	15.07
Accounting for the censoring until 1995, θ	0.03	0.01	0.039
Shift in the probability of staying in the first year due to early attrition, δ	–3.82	0.49	0.022
Personal discount factor, β	3.14	0.10	0.96
Log likelihood	11,743		
Sample size	6,023		

figure just below the 1992 estimate. The changes for the 1997 and 1998 are not statistically significantly different from 0, implying that mean tastes were about $14,000 between 1996 and 1998. However, we find a statistically significant increase in mean tastes of $3,274 for the 1999 entry cohort, to –$11,028, but no statistically significant change for the 2000 entry cohort.

In short, we find that mean tastes decrease from around –$14,000 for the 1992 entry cohort to –$17,000 for the 1993–1995 entry cohorts, increase back to about –$14,000 for the 1996–1998 entry cohorts, and then further increase to –$11,000 for the 1999–2000 entry cohorts. These results suggest that mean tastes vary across the entry cohorts for a subgroup defined to be homogeneous.

It is unclear how important these differences are. In Figure 5.4, we find that responsiveness to a 1-percent across-the-board real pay decrease varies between –5.3 percent for the 1992 entry cohort and –3.2 percent for the 2000 entry cohort, in models that do not restrict the sample to be homogeneous like we do for the results in Table 5.5. Thus, we find some differences in pay responsiveness.

Model Selection for Conducting Policy Simulations

To conduct the policy simulations presented in Chapter Six, we need to select a set of estimates. Which entry cohort should we use for conducting policy simulations? Given the evidence that estimated mean taste differs across entry cohorts and pay responsiveness differs as well, the selection can affect the policy simulation results. One might argue for the use of the 2000 entry cohort as the most recent entry cohort, but the results of earlier entry cohorts might also be relevant. For that reason, we conduct the policy simulations using a model that combines entry cohorts. Given the better fit of the data with the model that combines the 1992–2000 entry cohorts, we use the 1992–2000 combined model estimates, shown in the right panel of Table 5.4, to conduct policy simulations.

To get a sense of how simulations with the combined model compare with the separate models, we can compare the simulated changes in steady-state force size as a result of a 1-percent decrease in pay. For the combined model, we find a –3.7-percent drop in steady-state force size (Table 5.4). For the separate models, we find a –3.2-percent drop for the 2000 entry cohort and a –5.3-percent drop for the 1992 entry cohort. Thus, the combined model result falls within the range of the separate models.

CHAPTER SIX

Policy Simulations

The model estimates presented in Chapter Five permit us to simulate how changes in compensation policy affect DoD civilian employee retention. We now illustrate this simulation capability by analyzing the effect that the higher employee contribution rates mandated under Pub. L. 112-96 have on DoD civilian retention. Employees hired in 2013 had to contribute 3.1 percent of their salaries into the Civil Service Retirement and Disability Fund, up from 0.8 percent for those hired prior to that point. Employees hired in 2014 and later are required to contribute 4.4 percent (Zawodny, 2013). These contributions help cover the cost of the FERS defined benefit plan or basic plan.

We chose this mandated increase in employee contributions to illustrate the simulation capability for three reasons. First, little information is available on the effects of the policy, given that it is relatively recent, and understanding its effects is particularly important given that some in Congress seek to further increase mandated employee contributions (Lunney, 2014). Second, because we have already conducted a preliminary analysis of the retention effects of Pub. L. 112-96 using our earlier prototype model based on the 1988 entry cohort (Asch, Mattock, and Hosek, 2014b), we have the opportunity to compare the simulation results from our updated model with our earlier results. Finally, simulations of the effects of Pub. L. 112-96 allow us to highlight the new costing capability we have developed because the law changes DoD retirement costs by changing DoD's contributions. Specifically, OPM estimates that the cost of the FERS defined benefit plan is 12.7 percent of pay (Isaacs, 2014). Thus, when employees contribute 0.8 percent, the government contribution is 11.9 percent (12.7 – 0.8); when they contribute 3.1 percent, the government contribution falls to 9.6 percent (12.7 – 3.1).

In this chapter, we show the steady-state retention and cost effects of Pub. L. 112-96, using the DRM estimates for the model that combines the 1992–2000 entry cohorts, shown in Table 5.4 in Chapter Five. The results we present assume that hiring is unchanged, so we assume that hiring does not offset changes in retention.

Accounting for Savings Behavior

The effect that higher mandated employee contribution rates have on retention depends on employees' saving behavior. Employees can choose how much of their current pay to spend or save. If employees were already saving enough of their base pay to cover the mandated increase in FERS contributions, or at least 4.4 percent of their pay, the amount of money from each paycheck available for current expenditures would stay the same.[1] In some cases, employees might choose to shift contributions from one part of the FERS plan to another—lowering the amount deposited into the FERS TSP and using that money to cover the higher contributions to the defined benefit plan. When employees lower their TSP contributions, the amount that their agencies contribute to the TSP is also reduced. This is a costly change to the employee because shifting $1.00 out of the TSP means up to $2.00 less saved in the TSP, the employee's $1.00 plus the $1.00 match by the government. The end result is that FERS retirement benefits will be smaller with the higher mandated contribution rate.

In other cases, employees might cover the increased contribution by shifting money from other types of savings, such as stocks or bonds, leaving TSP contributions unchanged. Of course, if employees were saving less than 4.4 percent, they will have to save more out of their paychecks to reach the mandated contribution, which means they will have less money to spend today.

Alternatively, employees might not choose to cover the increased contributions by shifting money from TSP contributions or from other types of savings, possibly because their savings and TSP contributions are too small to cover the mandated FERS contribution, or because they do not want to give up the TSP match or do not want to decrease the financial liquidity provided by their savings account. These individuals, instead, might choose to cover the higher FERS contributions by lowering the money from each paycheck available to be spent on current consumption.

We have not yet incorporated into the DRM how individuals make saving decisions and, therefore, how those decisions change when they face a policy change, such as Pub. L. 112-96. Therefore, to assess the effects of the policy change, we consider a range of alternative assumptions about saving decisions that allow us to bound the range of effects that the higher retirement contribution could have on retention. Specifically, we consider three cases:

1. Employees reduce other savings, leaving TSP contributions and take-home pay unchanged.
2. Employees reduce their TSP contributions, reducing the TSP match they receive from DoD from 5 percent to 3 percent.
3. Employees lower current consumption, so the earnings available for spending are lower. We model this case as a reduction in current federal pay. Given that the mandated contribution rate increased from 0.8 percent of salary to 4.4 percent, the decrease in salary is 3.6 percent.

---

[1]   It is possible that the shifting of funds away from other investments toward contributions to the defined benefit portion of FERS could cause individuals to reoptimize and even increase their savings to ensure achieving a given targeted saving level at retirement. It is also possible that wealth, including savings, over the lifetime of the member is lower when FERS contributions increase because members who were saving—either in FERS or in some other investment vehicle—with the expectation of a future return are now forced to provide higher FERS contributions without a change in the FERS returns. Changes in overall wealth, including savings, could affect retention in the civil service. Because we do not model the saving decision, we cannot incorporate these possible effects on retention behavior. That said, as discussed in Chapter Five, our model incorporates both current compensation and future retirement wealth and fits the data well.

Results

Figures 6.1 through 6.3 and Table 6.1 summarize the simulation results for three cases. The figures show the steady-state effect on DoD employee retention, while the table shows the steady-state force size and personnel costs in the baseline and under the policy change for the three cases. The last column in the table shows the percentage change in force size, as well as the percentage change in cost per person. We consider the change in cost per person as a means of holding constant force size under the policy change (although not the experience mix of the force). That is, we do not incorporate policies to sustain the overall size of the force, such as a retention bonus, into the measures of total cost, so our discussion of how the different cases affect cost focuses on the change in cost per person. To scale the simulations, we use the number of full-time, nonseasonal employees under the GS pay plan with at least bachelor's degrees at the end of 2011, computed using Defense Manpower Data Center civilian personnel data for DoD. We find that there were 225,888 such employees in 2011, and we use this figure to scale the simulations.[2] The choice of 2011 as a baseline is somewhat arbitrary, and alternatives could, of course, be considered. Costs are measured in billions of 2015 dollars.

Consider case 1 first. If employees cover the increase in mandated contributions by shifting money from other types of savings, such as stocks or bonds, leaving TSP contributions and current compensation available for consumption unchanged, we find no change in retention (Figure 6.1). As seen in Table 6.1, force size remains unchanged, as do salary costs. The cost of the retirement plan is the accrual charge associated with the FERS basic plan and special retirement supplement that includes both the DoD and employee contributions. The effect of the retirement plan's cost on DoD is the portion of the accrual charge that DoD pays to the Civil Service Retirement and Disability Fund, i.e., the total retirement plan cost minus employee contributions. Both are intragovernmental transfers. Because there is no change in the size or experience mix of the force, the total cost of the retirement plan and, therefore, total cost are unchanged. But, DoD's retirement costs decline as employees bear a larger share of the contributions as a result of the mandated increase in their contributions. In the steady state, we find that DoD retirement costs per person fall by more than 18 percent, while DoD's total costs per person fall by 3 percent.

Next, consider case 2 in which we assume that employees cover the higher mandated contributions by reducing their TSP contributions, thereby reducing the TSP match rate to 3 percent from 5 percent. We find that the size of the steady-state GS workforce that is retained falls by 3.7 percent. As shown in Figure 6.2, the fall occurs among those with fewer than 30 years of service, so the force becomes more senior on average. Although force size falls, cost per person only falls by 0.2 percent (Table 6.1). Retirement costs per person decrease because DoD's TSP match rate decreases, reducing the cost of TSP contributions, and because the force is smaller, so the salary bill is lower and hence so is the accrual charge. DoD retirement

---

[2]   We use the Defense Manpower Data Center GS and equivalent or broadly defined GS pay plan categorization that includes the GS, CZ, GG, GW, and GM codes. Importantly, the 225,888 employees include those under all retirement plans, not just FERS. The reason we did not restrict our tabulations to those only under FERS is that FERS became effective January 1, 1987, and has not been existence long enough for an employee to have spent an entire career under this system. For our analysis, we require an estimate of the steady-state number of GS employees with at least bachelor's degrees under FERS. Had we restricted the tabulation to include only those under FERS, we would have produced an undercount of the steady-state number of employees. We estimate the steady-state number under FERS by using the size of the DoD GS workforce with at least bachelor's degrees in 2011, including employees under all retirement systems.

**Figure 6.1
Steady-State Effects of Higher Mandated Employee
Contributions with Lower Non–Thrift Savings Plan
Savings: Case 1**

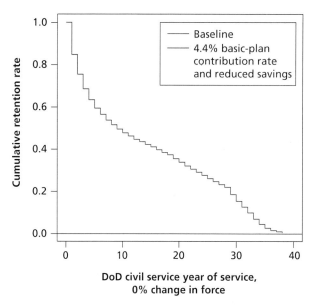

DoD civil service year of service,
0% change in force

NOTE: The baseline retention curve perfectly overlaps with
the policy simulation. This is all education levels for a DoD
GS simulation.

RAND *RR1503-6.1*

**Figure 6.2
Steady-State Effects of Higher Mandated Employee
Contributions with Lower Thrift Savings Plan
Contributions: Case 2**

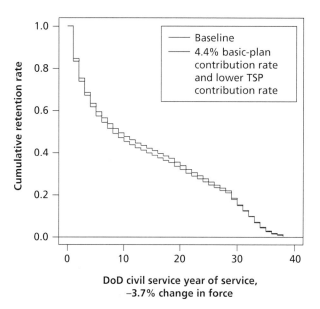

DoD civil service year of service,
−3.7% change in force

NOTE: This is all education levels for a DoD GS simulation.

RAND *RR1503-6.2*

Figure 6.3
Steady-State Effects of Higher Mandated Employee
Contributions with Lower Current Consumption and
Lower Take-Home Pay: Case 3

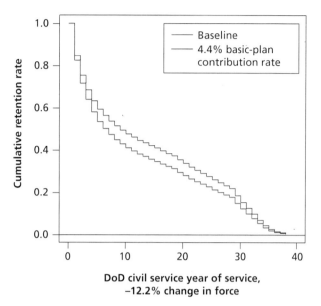

NOTE: This is all education levels for a DoD GS simulation.
RAND *RR1503-6.3*

costs further decrease because its share of the retirement plan's costs decreases when mandated employee contributions increase (assuming a constant overall accrual charge). We find that DoD's retirement cost per person falls by nearly 30 percent, while DoD's total cost per person falls by 5 percent.

Finally, under case 3, we consider the retention and cost effects when employees reduce consumption by reducing the amount of money from each paycheck available to be spent. We model this change as being equivalent to a pay cut of 3.6 percent (4.4 – 0.8). Figure 6.3 shows that such a cut in pay would reduce the size of the steady-state workforce that is retained by 12.2 percent. Furthermore, the reduction in the steady state would occur at all years of service, even beyond the 30th year of service, although the steady-state reductions are larger in early years of service leading to lower retention by the midcareer years (i.e., ten to 30 years of service). Salary costs per person fall by 1.1 percent because employees are more junior on average, while retirement costs per person fall by 2.2 percent. The total cost to DoD per person falls by 1.3 percent.

Comparisons

As mentioned earlier in the chapter, we chose the three cases to bound the range of possible results. The smallest changes occur under case 1, in which we find no change in retention or government cost per person and a 3-percent reduction in DoD cost per person. The largest change in retention occurs under case 3—a 12.2-percent reduction in force size—in which we assume that individuals finance the higher mandated contribute by consuming less, equivalent

Table 6.1
Simulated Effects of Higher Mandated Employee Contributions on General Schedule Force Size and Personnel Costs, Three Cases

Case or Effect	Baseline	Policy	Percentage Change in Retention, or Cost per Person from Baseline
Case 1: Reduced (non-TSP) savings			
Force size, in number of personnel	225,888	225,888	0.0
Salary cost, in billions of 2015 dollars	16.73	16.73	0.0
Retirement plan cost, in billions of 2015 dollars	3.26	3.26	0.0
DoD retirement plan cost, in billions of 2015 dollars	3.26	2.65	−18.5
DoD total cost, in billions of 2015 dollars	19.99	19.38	−3.0
Case 2: Reduced TSP contributions			
Force size, in number of personnel	225,888	217,428	−3.7
Salary cost, in billions of 2015 dollars	16.73	16.07	−0.2
Retirement plan cost, in billions of 2015 dollars	3.26	2.79	−10.9
DoD retirement plan cost, in billions of 2015 dollars	3.26	2.21	−29.4
DoD total cost, in billions of 2015 dollars	19.99	18.28	−5.0
Case 3: Reduced consumption (take-home pay)			
Force size, in number of personnel	225,888	198,267	−12.2
Salary cost, in billions of 2015 dollars	16.73	14.53	−1.1
Retirement plan cost, in billions of 2015 dollars	3.26	2.79	−2.2
DoD retirement plan cost, in billions of 2015 dollars	3.26	2.27	−20.6
DoD total cost, in billions of 2015 dollars	19.99	16.80	−4.2

NOTE: "DoD retirement plan cost" and "DoD total cost" exclude mandated member contributions to the FERS basic plan, while "Retirement plan cost" includes these contributions, thereby reflecting the full personnel costs from current and deferred compensation. Retirement costs include the accrual costs associated with the FERS basic plan and the special retirement supplement plus DoD TSP contribution costs. Total cost is the sum of salary costs and retired pay costs.

to a reduction in salary available for spending. However, in terms of changes in cost per person, we find that the largest changes occur under case 2, in which we assume that employees finance the higher mandated contributions with lower TSP contributions that, in turn, result in lower DoD contributions. DoD costs per person fall by 5 percent. That said, the changes in cost per person under case 3 are not much less than under case 2, a 4.2-percent drop in DoD cost per person. In case 3, most of the reduction in DoD cost per person is due to reduced salary costs per person (because the workforce is more junior on average), while, in case 2, most of the reduction is due to lower DoD retirement costs per person. Overall, the results suggest a range of decrease in DoD current and deferred compensation costs per person of between 3 percent and 5 percent.

Finally, we can compare our results with the results from our earlier analysis using our prototype model based on the retention behavior of the 1988 entry cohort. The earlier analysis, summarized in Asch, Mattock, and Hosek, 2014b, also considered the three cases above, although it analyzed only the retention and not the cost effects of these cases. Like in the current analysis, the earlier analysis found no change in retention under case 1. The largest retention change was under case 3, also like in the current analysis. That said, the change in retention under case 3 was less, a reduction of 8.6 percent, rather than 12.2 percent in the current analysis. The more-modest responsiveness to the policy change found in the earlier study is not surprising, given our simulation results of a 1-percent real pay drop shown in Tables 5.3 and 5.4 in Chapter Five. In Table 5.4, we find a 3.74-percent drop in the workforce that is retained using the model estimates that combine the 1992–2000 entry cohorts (and the basis for the simulations in this chapter). In Table 5.3, we find a 2.1-percent drop using estimates based on the 1988 entry cohort. Thus, our updated results are consistent with our earlier analysis, although they suggest somewhat greater retention responsiveness to the policy change.

CHAPTER SEVEN

Conclusions

New Capability for Policy Analysis

The chief objective of this research was to develop greater capability for determining the retention and costs of changes in DoD civilian compensation policy. The earlier, prototype DRM, which provided a starting point, was estimated with data on a single entry cohort of DoD civilian workers, entrants in 1988, and had no costing capability. The current research extends and enhances the prototype. The retention database now includes entering entry cohorts from 1988 to 2000. We have estimated new pay profiles from ACS data, and a comparison to pay profiles that we generated from synthetic birth cohorts constructed from CPS data match the ACS profiles well. Also, we estimated DRMs by entry cohort. We found some differences in parameter estimates, but they were not large and led us to choose a DRM estimated on a combined sample and specified without entry cohort–specific interactions, for use in policy simulation. Finally, the policy simulations in Chapter Six not only showed the retention effects but also the policy costs relative to those at baseline. Overall, the wider range of entry cohort retention data, refined pay profiles, and assessment of parameters by entry cohort versus the combined model provide a stronger basis to recommend the DRM for the analysis of DoD civilian compensation policy, and the addition of costing provides an important element for the comparison of policies.

Model Extensions

Additional extensions would continue to improve the DRM capability, and these should be considered as areas of fruitful future research. First, the DRM capability can be used to simulate the retention effects of other policies of interest. Indeed, we are using the extended model to assess the cost-effectiveness of severance pay and voluntary separation incentives as policies to achieve a given drawdown or workforce-shaping goal. The DRM capability could also be extended to other occupational areas within DoD, including the cyber workforce; to other pay systems, such as the science, technology, engineering, and math workforces in the various demonstration programs and the wage grade workforce; to specific demographic groups, such as women and minorities; and to specific locations of interest, such as Hawaii. Furthermore, with appropriate data, the DRM capability could be applied to civil service workforces in other agencies within the federal government, including the U.S. Department of Veterans Affairs, the U.S. Department of Homeland Security, and the various agencies that make up the intelligence community.

The DRM itself could be extended in important ways in future research. First, we could incorporate into the model the fact that some civil service personnel leave and then return. Returnee behavior has already been introduced in models of reserve-component participation in the military, and it can be introduced in the civil service model as well. Second, we could incorporate changes in expectations about future policy changes. Pay freezes and furloughs can lead to changed expectations about the likelihood of future pay freezes and furloughs. Workers might view future pay as more uncertain. Such an extension of the model would require incorporating a model of how workers form and change their expectations.

Third, the model could be extended to create a "total force" model of DoD workforce dynamics and the effects that compensation has on those dynamics, in which *total force* includes active and reserve military personnel and DoD civilians (but not contractors). As shown in Gates et al., 2008, about half of new DoD civil service hires have prior military service. Furthermore, many DoD civilians also participate in the reserve components as drilling selected reservists. Consequently, changes in compensation and personnel policy in the active component, the reserve component, or the federal civil service could affect retention in all three parts of the DoD workforce in interrelated ways. RAND has already developed a unified DRM capability to provide logically consistent and empirically based estimates of the effects that compensation policy has on active-component retention and reserve-component participation. This capability could be extended to include DoD civil service employment.

Estimation of Pay Profiles

We use an upper-censored Tobit model to estimate the profiles of federal civil service pay and private-sector pay. We also consider a model that accounts for differences in entry cohort–specific pay profiles. In the first subsection, we present the basic specification for the Tobit model; in the following subsection, we discuss an extension permitting variation in pay profiles by entry cohort.

Tobit Model Using Cross-Sectional Data

A Tobit model assumes a linear specification, $y = X\beta + \varepsilon$, where y represents continuous outcomes and the error is normally distributed and independent across observations, $\varepsilon \sim N(0, \sigma^2 I)$. We observe individual i's log earnings, y_i, for observations $i \in C$. Observations $i \in R$ are right-censored; we know only that they are greater than or equal to the known threshold y_{Ri}. The log likelihood is

$$\ln L = -\frac{1}{2}\Sigma_{i \in C} w_i \left\{ \left(\frac{y_i - X_i\beta}{\sigma} \right)^2 + \log(2\pi\sigma) \right\} + \Sigma_{i \in R} w_i \log\left\{ 1 - \phi\left(\frac{y_{Ri} - X_i\beta}{\sigma} \right) \right\},$$

(A.1)

where $\phi(\ )$ is the standard cumulative normal and w_i is the weight for the ith observation.

We estimate the model on two populations: male federal civil service workers ages 25 to 65 and male private-sector wage and salary workers ages 25 to 65. The dependent variable, y, is the natural logarithm of wage and salary income in 2013 dollars. For each population, the model controls for education using indicator variables for at least a master's degree, at least a professional degree, and at least a doctoral degree (left out group: bachelor's only), veteran status, an indicator for whether the person served on active duty between August 1990 and August 2001 and after August 2001, year indicators, and an age spline.

As seen in Figure 4.1 in Chapter Four, pay profiles increase sharply at first, before moderately declining after age 50. A quadratic in age would have difficulty fitting this pattern. As an alternative, we use an age spline (i.e., a piecewise linear function), in which the knots (or points at which the spline changes slope) depend on age. $Age_a(age_i)$, $a = 1, \ldots, K$ represents K variables to be created and k_a, $a = 1, \ldots, K-1$ are the corresponding knots. We set knots at five-year age groups between 25 and 65, although the last age group is six years long, from 60 to 65:

$$Age_a(\ ) = age_i \qquad\qquad \text{if } a = 1$$
$$Age_a(\ ) = \max(0, age_i - k_{a-1}) \quad \text{if } a = 2 \ldots K.$$

The specification for $X_i\beta$ is

$$\beta_{intercept} + \beta_{a=1}Age_{a=1}(\ ) + \beta_2 Age_2(\ ) + \cdots + \beta_K Age_K(\ ) + \beta_{MA}masters.degree_i$$
$$+ \beta_{PRO}\,professional.degree_i + \beta_{DOC}doctoral.degree_i + \beta_{VET}veteran_i + \beta_{VET90}veteran90_i$$
$$+ \beta_{VET01}veteran01_i + \sum_{\substack{y=[2003,2012] \\ y \neq 2011}} \left\{ \beta_y I\left[year_i = y \right] \right\}.$$

In the case in which the model is estimated on the ACS, an additional set of indicator variables is added to account for the variation in the state top-code. We organized states into seven groups based on their top-coding thresholds.

The log-transformed pay profile for a male with a bachelor's degree who is not a veteran and faces 2011 aggregate conditions is determined by

$$y = \begin{cases} \beta_{intercept} + \beta_{a=1}Age_{a=1} & age_i \in [25,29] \\ \beta_{intercept} + \beta_{a=1}Age_{a=1} + \beta_2 Age_2 & age_i \in [30,34] \\ \quad\vdots & \quad\vdots \\ \beta_{intercept} + \beta_{a=1}Age_{a=1} + \beta_2 Age_2 + \cdots + \beta_8 Age_8 & age_i \in [60,65] \end{cases}.$$

The marginal effect is

$$\frac{dy}{dage} = \begin{cases} \beta_{a=1} & age_i \in [25,29] \\ \beta_{a=1} + \beta_{a=2} & age_i \in [30,34] \\ \quad\vdots & \quad\vdots \\ \beta_{a=1} + \beta_{a=2} + \cdots + \beta_{a=8} & age_i \in [60,65] \end{cases}.$$

Table A.1 presents the results. We observe that pay growth is greatest in the late 20s, with the salary trajectory tapering off in the 30s, and flat by the early 50s (note that the marginal effect of a spline is cumulative, e.g., for model 1, the marginal effect of one year of age on log pay is $0.057 - 0.028 = 0.029$). The reduction in pay growth is less gradual for private-sector workers, which could be partially due to a lower baseline pay. The permanent returns to education are greater in the private sector and greater in the CPS sample than in the ACS sample. The ACS finds a negative association between veteran status and pay in the federal sector, but this association does not exist in the CPS. A negative association with pay and veteran status in the private sector is found in the ACS and CPS. These negative associations could be due to

Table A.1
Tobit Regression for Samples from the American Community Survey and the Current Population Survey, 2003 Through 2012

	Federal		Private	
Variable	(1) ACS	(2) CPS	(3) ACS	(4) CPS
Age, knot 1 (25–29)	0.0573***	0.0489***	0.0603***	0.0529***
	(0.00457)	(0.0116)	(0.00104)	(0.00368)
Age, knot 2 (30–34)	−0.0284***	−0.0332*	−0.0126***	−0.00788
	(0.00667)	(0.0182)	(0.00167)	(0.00586)
Age, knot 3 (35–39)	−0.00314	0.0179	−0.0239***	−0.0194***
	(0.00499)	(0.0138)	(0.00152)	(0.00519)
Age, knot 4 (40–44)	−0.0124***	−0.0229**	−0.0140***	−0.0205***
	(0.00450)	(0.0113)	(0.00155)	(0.00534)
Age, knot 5 (45–49)	−0.0124***	−0.00846	−0.0100***	−0.00400
	(0.00436)	(0.0112)	(0.00162)	(0.00580)
Age, knot 6 (50–54)	−0.00204	−0.000152	−0.0075***	−0.0121*
	(0.00452)	(0.0127)	(0.00169)	(0.00638)
Age, knot 7 (55–59)	−0.000252	−0.00213	−0.00393**	0.000865
	(0.00426)	(0.0139)	(0.00186)	(0.00759)
Age, knot 8 (60–65)	0.00287	−0.00955	−0.00152	−0.00143
	(0.00543)	(0.0192)	(0.00252)	(0.0111)
Master's degree	0.154***	0.222***	0.194***	0.230***
	(0.00543)	(0.0157)	(0.00194)	(0.00710)
Professional degree	0.417***	0.464***	0.493***	0.656***
	(0.00889)	(0.0318)	(0.00395)	(0.0174)
Doctoral degree	0.260***	0.303***	0.269***	0.423***
	(0.00912)	(0.0268)	(0.00426)	(0.0158)
Year indicator: 2004	0.0409***	0.0549*	0.0131**	−0.0297**
	(0.0147)	(0.0298)	(0.00519)	(0.0131)
Year indicator: 2005	0.0295***	0.0559*	−0.0102**	−0.0274**
	(0.0112)	(0.0290)	(0.00437)	(0.0133)
Year indicator: 2006	0.0140	0.0258	−0.0230***	−0.0171
	(0.0110)	(0.0277)	(0.00426)	(0.0133)
Year indicator: 2007	0.0319***	0.0587**	−0.00962**	−0.0220
	(0.0109)	(0.0297)	(0.00426)	(0.0136)
Year indicator: 2008	0.0152	0.0542*	−0.0399***	−0.0199

Table A.1—Continued

Variable	Federal		Private	
	(1) ACS	(2) CPS	(3) ACS	(4) CPS
	(0.0111)	(0.0280)	(0.00427)	(0.0128)
Year indicator: 2009	0.0568***	0.0599**	0.00141	−0.0116
	(0.0110)	(0.0297)	(0.00424)	(0.0128)
Year indicator: 2010	0.0512***	0.0826***	−0.0168***	−0.0309**
	(0.0110)	(0.0302)	(0.00426)	(0.0134)
Year indicator: 2011	0.0439***	0.0673**	−0.0323***	−0.0524***
	(0.0114)	(0.0296)	(0.00442)	(0.0128)
Year indicator: 2012	0.0242**	0.0596**	−0.0422***	−0.0487***
	(0.0112)	(0.0301)	(0.00431)	(0.0130)
Veteran	−0.0598***	−0.00580	−0.0331***	−0.0381***
	(0.00697)	(0.0185)	(0.00357)	(0.0124)
Veteran (last service 1990–2000)	0.00677	0.0413	−0.00157	−0.0329
	(0.00940)	(0.0501)	(0.00592)	(0.0242)
Veteran (last service after 2000)	0.00197	−0.0312	0.00550	0.0263
	(0.00927)	(0.0483)	(0.00747)	(0.0384)
ACS top-coding threshold, tier 2 ($190,000–200,000)	0.000632		0.0632***	
(States: Alaska, Ariz., Fla., Kan., La., Mich., Mo., Neb., N.C., Ohio, Ore., R.I., Tenn., Utah, Vt., and Wis.)	(0.00760)		(0.00274)	
ACS top-coding threshold, tier 3 ($210,000–227,000)	0.0338***		0.152***	
(States: Del., Ga., N.H., N.D., Texas, Pa., and Wash.)	(0.00824)		(0.00298)	
ACS top-coding threshold, tier 4 ($250,000)	0.168***		0.205***	
(States: Calif., Colo., Ill., Md., Minn., and Va.)	(0.00717)		(0.00281)	
ACS top-coding threshold, tier 5 ($300,000)	0.0678***		0.241***	
(States: Mass., N.J., and N.Y.)	(0.0102)		(0.00325)	
ACS top-coding threshold, tier 6 ($355,000)	0.0279		0.321***	
(State: Conn.)	(0.0302)		(0.00763)	
ACS top-coding threshold, tier 7 ($375,000)	0.236***		0.260***	
(State equivalent: D.C.)	(0.0138)		(0.0120)	
Constant	9.173***	9.494***	9.000***	9.397***
	(0.130)	(0.327)	(0.0299)	(0.104)
Observations	14,146,097	1,075,473	14,294,179	1,075,473

NOTE: SEs are in parentheses. *** $p < 0.01$. ** $p < 0.05$. * $p < 0.1$.

the nature of work that veterans pursue in these sectors, which might differ from observationally similar people. Within the ACS sample, the indicators for top-coding group are generally as expected in the private sector: As the top-coding threshold increases, so does the pay level in that state. This is not true in the federal civil service sector. This is likely due to the composition and location of jobs. For example, if federal civil service jobs are disproportionately located in rural areas (e.g., military posts), federal pay might appear to be lower than others in a state. Year effects differ substantially between the federal and private sectors, and these differences were reflected in both the ACS and CPS samples. Relative to 2003, the federal-sector pay was generally higher, while the opposite was true in the private sector. Overall, the results in Table A.1 suggest that we achieve similar outcomes using the ACS or the CPS samples, with the exception of the returns to education and veteran status in the federal civil service noted above.

For pay-profile predictions from the estimated model, we assume a male with a bachelor's degree who is not a veteran, in year 2011. When we predicted the pay profile based on the model estimated using the ACS data, we used the observed states for each person. The resulting pay profiles, presented in Figure 4.3 in Chapter Four, reflect the average prediction of the weighted sample.

Tobit Model Accounting for Variation in Pay Profiles by Birth Cohort

In this section, we extend the model in Equation A.1 to account for differences by birth cohort. Each individual i is designated as belonging to a birth cohort defined by birth year; a member of birth cohort c was born in years $c \in [c-2, c+2]$. We define birth cohorts in five-year intervals—namely, 1943, 1948,...,1978, 1983—and set 1963 as the baseline birth cohort.

We use the same controls and a piecewise linear specification in age as in the model using cross-sectional data. The specification for $X_i\beta$ is

$$\beta_{intercept} + \beta_{a=1}Age_{a=1}(\ ) + \beta_2 Age_2(\ ) + \cdots + \beta_K Age_K(\ ) + \beta_{MA}masters.degree_i$$
$$+ \beta_{PRO}professional.degree_i + \beta_{DOC}doctoral.degree_i + \beta_{VET}veteran_i + \beta_{VET90}veteran90_i$$
$$+ \beta_{VET01}veteran01_i$$
$$+ \sum_{\substack{c \in \{COHORTS\} \\ c \neq \text{ baseline cohort}}} Cohort_i$$
$$\times \left\{ \begin{array}{l} \beta_{c,intercept} + \beta_{c,a=1}Age_{a=1}(\ ) + \beta_{c,2}Age_2(\ ) + \cdots + \beta_{c,K}Age_K(\ ) + \beta_{c,MA}masters.degree_i \\ + \beta_{c,PRO}professional.degree_i + \beta_{c,DOC}doctoral.degree_i + \beta_{c,VET}veteran_i + \beta_{c,VET90}veteran90_i \\ \beta_{c,VET01}veteran01_i \end{array} \right\}$$
$$+ \sum_{\substack{y=[1964,2014] \\ y \neq 2011}} \{\beta_y I[year_i = y]\}.$$

The age–earning profile for a male in the 1963 birth cohort with a bachelor's degree who is not a veteran and faces 2011 aggregate conditions is determined by

$$
y_{c=1963} = \begin{cases}
\beta_{intercept} + \beta_{a=1}Age_{a=1} & age_i \in [25,29] \\
\beta_{intercept} + \beta_{a=1}Age_{a=1} + \beta_2 Age_2 & age_i \in [30,34] \\
\vdots & \vdots \\
\beta_{intercept} + \beta_{a=1}Age_{a=1} + \beta_2 Age_2 + \cdots + \beta_8 Age_8 & age_i \in [60,65]
\end{cases}.
$$

For nonbaseline birth cohorts, it is determined by

$$
y_{c \neq 1963} = \begin{cases}
y_{c=1963} + \beta_{c,intercept} + \beta_{c,a=1}Age_{a=1} & age_i \in [25,29] \\
y_{c=1963} + \beta_{c,intercept} + \beta_{c,a=1}Age_{a=1} + \beta_{c,2} Age_2 & age_i \in [30,34] \\
\vdots & \vdots \\
y_{c=1963} + \beta_{c,intercept} + \beta_{c,a=1}Age_{a=1} + \beta_{c,2} Age_2 + \cdots + \beta_{c,8} Age_8 & age_i \in [60,65]
\end{cases}.
$$

The marginal effect of age for the 1963 baseline birth cohort is

$$
\frac{dy_{c=1963}}{dage} = \begin{cases}
\beta_{a=1} & age_i \in [25,29] \\
\beta_{a=1} + \beta_{a=2} & age_i \in [30,34] \\
\vdots & \vdots \\
\beta_{a=1} + \beta_{a=2} + \cdots + \beta_{a=8} & age_i \in [60,65]
\end{cases},
$$

and the marginal effect of age for nonbaseline birth cohorts c is

$$
\frac{dy_{c \neq 1963}}{dage} = \begin{cases}
(\beta_{a=1} + \beta_{c,a=1}) & age_i \in [25,29] \\
(\beta_{a=1} + \beta_{c,a=1}) + (\beta_{a=2} + \beta_{c,a=2}) & age_i \in [30,34] \\
\vdots & \vdots \\
(\beta_{a=1} + \beta_{c,a=1}) + (\beta_{a=2} + \beta_{c,a=2}) + \cdots + (\beta_{a=8} + \beta_{c,a=8}) & age_i \in [60,65]
\end{cases}.
$$

Table A.2 presents the results of the model. It is important to note that 1963 is the baseline birth cohort. This implies that the main regression results, such as age, education, and veteran status, are based on the associations with the 1963 birth cohort. The results in Table A.2 can be hard to interpret for the age splines, given the cumulative nature of the marginal effects described above. As a result, we prefer to examine the model predictions, as presented in Figures 4.4 and 4.5 in Chapter Four, and refer to Table A.2 to understand deviations from the cross-sectional results. We simulated the birth cohort pay profiles using the model estimates. In generating the model's pay-profile predictions, we assumed that the male federal or private-sector worker had only a bachelor's degree, was not a veteran, and faced the aggregate conditions present in 2011.

Table A.2
Tobit Regression for Samples from the Current Population Survey, 1966 Through 2014, Controlling for Birth Cohort

Variable	(1) Federal	(2) Private
Age, knot 1 (25–29)	0.0492**	0.0603***
	(0.0239)	(0.00526)
Age, knot 2 (30–34)	−0.0277	−0.0249***
	(0.0352)	(0.00881)
Age, knot 3 (35–39)	−0.00804	−0.00706
	(0.0221)	(0.00923)
Age, knot 4 (40–44)	0.00493	−0.0284***
	(0.0216)	(0.00902)
Age, knot 5 (45–49)	−0.0233	0.00227
	(0.0214)	(0.00952)
Age, knot 6 (50–54)	0.0415	−0.0206
	(0.0365)	(0.0199)
Age, knot 7 (55–59)	0.0192	0.0291
	(0.0355)	(0.0252)
Age, knot 8 (60–65)	0.0132	−0.000413
	(0.0545)	(0.0294)
Master's degree	0.230***	0.224***
	(0.0263)	(0.0127)
Professional degree	0.368***	0.553***
	(0.0431)	(0.0282)
Doctoral degree	0.115	0.374***
	(0.148)	(0.0285)
Veteran	−0.0797**	−0.105***
	(0.0334)	(0.0202)
1943 birth cohort	−0.478	0.260
	(1.060)	(0.280)
1948 birth cohort	−0.896	−0.0724
	(0.843)	(0.226)
1953 birth cohort	−0.433	0.0267
	(0.813)	(0.213)
1958 birth cohort	−0.586	0.219
	(0.805)	(0.205)

Table A.2—Continued

Variable	(1) Federal	(2) Private
1968 birth cohort	0.411	0.0970
	(0.784)	(0.213)
1973 birth cohort	0.390	0.455**
	(0.872)	(0.224)
1978 birth cohort	−0.171	0.151
	(0.906)	(0.222)
1983 birth cohort	−0.673	0.402*
	(0.837)	(0.225)
1943 birth cohort × knot 1	0.0143	−0.0106
	(0.0368)	(0.00969)
1948 birth cohort × knot 1	0.0252	−9.32e-05
	(0.0299)	(0.00797)
1953 birth cohort × knot 1	0.0110	−0.00259
	(0.0292)	(0.00760)
1958 birth cohort × knot 1	0.0211	−0.00734
	(0.0290)	(0.00737)
1968 birth cohort × knot 1	−0.0140	−0.00423
	(0.0279)	(0.00767)
1973 birth cohort × knot 1	−0.0165	−0.0163**
	(0.0310)	(0.00800)
1978 birth cohort × knot 1	0.00338	−0.00660
	(0.0319)	(0.00789)
1983 birth cohort × knot 1	0.0228	−0.0166**
	(0.0299)	(0.00802)
1943 birth cohort × knot 2	−0.0215	0.0164
	(0.0513)	(0.0141)
1948 birth cohort × knot 2	0.00368	0.00804
	(0.0429)	(0.0125)
1953 birth cohort × knot 2	0.0206	0.0127
	(0.0426)	(0.0123)
1958 birth cohort × knot 2	−0.0111	0.0160
	(0.0457)	(0.0128)
1968 birth cohort × knot 2	0.0249	0.0148

Table A.2—Continued

Variable	(1) Federal	(2) Private
	(0.0428)	(0.0131)
1973 birth cohort × knot 2	0.0255	0.0190
	(0.0461)	(0.0128)
1978 birth cohort × knot 2	−0.0229	0.00674
	(0.0466)	(0.0131)
1983 birth cohort × knot 2	−0.0196	0.0406*
	(0.0612)	(0.0214)
1943 birth cohort × knot 3	0.0635	−0.00858
	(0.0475)	(0.0135)
1948 birth cohort × knot 3	0.00200	−0.00465
	(0.0317)	(0.0125)
1953 birth cohort × knot 3	−0.0249	−0.0214*
	(0.0289)	(0.0126)
1958 birth cohort × knot 3	−0.0306	−0.0206
	(0.0313)	(0.0134)
1968 birth cohort × knot 3	−0.0128	−0.0235*
	(0.0305)	(0.0129)
1973 birth cohort × knot 3	0.00695	−0.0164
	(0.0321)	(0.0126)
1978 birth cohort × knot 3	0.103**	−0.00621
	(0.0449)	(0.0209)
1943 birth cohort × knot 4	−0.0750*	0.0176
	(0.0400)	(0.0141)
1948 birth cohort × knot 4	−0.0434	−0.00230
	(0.0289)	(0.0126)
1953 birth cohort × knot 4	−0.0324	0.0150
	(0.0273)	(0.0129)
1958 birth cohort × knot 4	0.0133	0.0121
	(0.0296)	(0.0136)
1968 birth cohort × knot 4	−0.0173	0.0238*
	(0.0285)	(0.0131)
1973 birth cohort × knot 4	−0.0547	0.0242
	(0.0421)	(0.0196)

Table A.2—Continued

Variable	(1) Federal	(2) Private
1943 birth cohort × knot 5	0.0344	−0.0215
	(0.0300)	(0.0154)
1948 birth cohort × knot 5	0.0162	−0.00440
	(0.0284)	(0.0135)
1953 birth cohort × knot 5	0.0342	0.00234
	(0.0280)	(0.0138)
1958 birth cohort × knot 5	0.0250	0.000442
	(0.0279)	(0.0140)
1968 birth cohort × knot 5	0.0100	−0.0178
	(0.0425)	(0.0214)
1943 birth cohort × knot 6	−0.0558	0.0227
	(0.0421)	(0.0240)
1948 birth cohort × knot 6	−0.0641	0.0107
	(0.0426)	(0.0228)
1953 birth cohort × knot 6	−0.0400	−0.000336
	(0.0426)	(0.0224)
1958 birth cohort × knot 6	−0.0670	0.0169
	(0.0414)	(0.0227)
1943 birth cohort × knot 7	−0.0212	−0.0532*
	(0.0449)	(0.0294)
1948 birth cohort × knot 7	0.00366	−0.0256
	(0.0419)	(0.0281)
1953 birth cohort × knot 7	−0.0381	−0.0275
	(0.0428)	(0.0277)
1943 birth cohort × knot 8	−0.0286	0.0114
	(0.0648)	(0.0351)
1948 birth cohort × knot 8	−0.0573	−0.00333
	(0.0659)	(0.0332)
1943 birth cohort × master's degree	−0.0326	−0.0105
	(0.0488)	(0.0280)
1948 birth cohort × master's degree	0.00102	−0.0140
	(0.0390)	(0.0210)
1953 birth cohort × master's degree	−0.0432	−0.0153

Table A.2—Continued

Variable	(1) Federal	(2) Private
	(0.0367)	(0.0188)
1958 birth cohort × master's degree	−0.0205	−0.0100
	(0.0339)	(0.0184)
1968 birth cohort × master's degree	−0.0563	0.0160
	(0.0349)	(0.0182)
1973 birth cohort × master's degree	−0.0941**	0.0212
	(0.0423)	(0.0184)
1978 birth cohort × master's degree	−0.0218	−0.0265
	(0.0535)	(0.0222)
1983 birth cohort × master's degree	−0.0106	0.0259
	(0.0960)	(0.0272)
1943 birth cohort × professional degree	−0.00406	0.0308
	(0.0750)	(0.0695)
1948 birth cohort × professional degree	0.0530	0.213***
	(0.0710)	(0.0446)
1953 birth cohort × professional degree	0.124**	0.0891**
	(0.0629)	(0.0444)
1958 birth cohort × professional degree	0.0898	0.122***
	(0.0701)	(0.0413)
1968 birth cohort × professional degree	0.0946	0.0864**
	(0.0722)	(0.0402)
1973 birth cohort × professional degree	−0.0323	−0.0303
	(0.0852)	(0.0444)
1978 birth cohort × professional degree	0.100	-0.108**
	(0.112)	(0.0506)
1983 birth cohort × professional degree	−0.0767	−0.209***
	(0.215)	(0.0559)
1943 birth cohort × doctoral degree	0.260*	0.0508
	(0.157)	(0.0507)
1948 birth cohort × doctoral degree	0.300*	0.0462
	(0.160)	(0.0423)
1953 birth cohort × doctoral degree	0.291*	0.0692*
	(0.153)	(0.0419)

Table A.2—Continued

Variable	(1) Federal	(2) Private
1958 birth cohort × doctoral degree	0.199	0.0195
	(0.154)	(0.0404)
1968 birth cohort × doctoral degree	0.0270	0.00765
	(0.157)	(0.0419)
1973 birth cohort × doctoral degree	−0.0381	−0.0421
	(0.168)	(0.0435)
1978 birth cohort × doctoral degree	−0.0190	0.00457
	(0.164)	(0.0504)
1983 birth cohort × doctoral degree	0.0726	−0.0318
	(0.169)	(0.0750)
1943 birth cohort × veteran	0.0416	0.138***
	(0.0408)	(0.0234)
1948 birth cohort × veteran	0.0428	0.106***
	(0.0379)	(0.0222)
1953 birth cohort × veteran	0.0341	0.0387
	(0.0393)	(0.0243)
1958 birth cohort × veteran	0.0175	−0.00549
	(0.0416)	(0.0275)
1968 birth cohort × veteran	0.0142	−0.00185
	(0.0454)	(0.0284)
1973 birth cohort × veteran	0.171***	0.0481
	(0.0580)	(0.0321)
1978 birth cohort × veteran	0.0641	−0.0102
	(0.0898)	(0.0409)
1983 birth cohort × veteran	0.0554	−0.00591
	(0.110)	(0.0616)
Year indicator: 1966	0.547***	0.133
	(0.189)	(0.0887)
Year indicator: 1967	0.165	0.119
	(0.183)	(0.0864)
Year indicator: 1968	0.168	0.112*
	(0.183)	(0.0656)
Year indicator: 1969	0.259	0.121*

Table A.2—Continued

Variable	(1) Federal	(2) Private
	(0.169)	(0.0620)
Year indicator: 1970	0.108	0.205***
	(0.156)	(0.0605)
Year indicator: 1971	0.203	0.201***
	(0.148)	(0.0568)
Year indicator: 1972	0.226	0.185***
	(0.145)	(0.0539)
Year indicator: 1973	0.276**	0.210***
	(0.140)	(0.0524)
Year indicator: 1974	0.367**	0.202***
	(0.145)	(0.0510)
Year indicator: 1975	0.224*	0.131***
	(0.136)	(0.0495)
Year indicator: 1976	0.182	0.131***
	(0.127)	(0.0479)
Year indicator: 1977	0.139	0.106**
	(0.125)	(0.0467)
Year indicator: 1978	0.147	0.0987**
	(0.119)	(0.0458)
Year indicator: 1979	0.0806	0.0889**
	(0.115)	(0.0449)
Year indicator: 1980	−0.0733	0.0695
	(0.110)	(0.0435)
Year indicator: 1981	−0.0249	0.0218
	(0.104)	(0.0422)
Year indicator: 1982	−0.0332	−0.0213
	(0.0992)	(0.0415)
Year indicator: 1983	−0.114	−0.0468
	(0.0966)	(0.0411)
Year indicator: 1984	−0.138	−0.0202
	(0.101)	(0.0401)
Year indicator: 1985	−0.128	−0.0187
	(0.0898)	(0.0388)

Table A.2—Continued

Variable	(1) Federal	(2) Private
Year indicator: 1986	−0.115	0.0119
	(0.0862)	(0.0377)
Year indicator: 1987	−0.154*	0.0252
	(0.0851)	(0.0371)
Year indicator: 1988	−0.195**	0.0293
	(0.0834)	(0.0361)
Year indicator: 1989	−0.163**	0.0123
	(0.0821)	(0.0355)
Year indicator: 1990	−0.151**	0.0135
	(0.0759)	(0.0343)
Year indicator: 1991	−0.259***	−0.0195
	(0.0807)	(0.0338)
Year indicator: 1992	−0.301***	−0.100***
	(0.0733)	(0.0328)
Year indicator: 1993	−0.271***	−0.101***
	(0.0699)	(0.0319)
Year indicator: 1994	−0.226***	−0.0979***
	(0.0682)	(0.0311)
Year indicator: 1995	−0.234***	−0.0856***
	(0.0671)	(0.0300)
Year indicator: 1996	−0.244***	−0.0742**
	(0.0653)	(0.0289)
Year indicator: 1997	−0.261***	−0.0989***
	(0.0631)	(0.0289)
Year indicator: 1998	−0.150**	−0.0544**
	(0.0616)	(0.0271)
Year indicator: 1999	−0.195***	0.0138
	(0.0594)	(0.0256)
Year indicator: 2000	−0.186***	0.0286
	(0.0575)	(0.0247)
Year indicator: 2001	−0.165***	0.0448*
	(0.0594)	(0.0230)
Year indicator: 2002	−0.146***	0.0506**

Table A.2—Continued

Variable	(1) Federal	(2) Private
	(0.0514)	(0.0222)
Year indicator: 2003	−0.110**	0.0316
	(0.0484)	(0.0207)
Year indicator: 2004	−0.0469	0.00479
	(0.0457)	(0.0197)
Year indicator: 2005	−0.0477	0.00991
	(0.0426)	(0.0184)
Year indicator: 2006	−0.0733*	0.0220
	(0.0390)	(0.0170)
Year indicator: 2007	−0.0408	0.0200
	(0.0380)	(0.0162)
Year indicator: 2008	−0.0378	0.0240*
	(0.0348)	(0.0146)
Year indicator: 2009	−0.0252	0.0349**
	(0.0345)	(0.0138)
Year indicator: 2010	0.00771	0.0173
	(0.0324)	(0.0134)
Year indicator: 2012	−0.00850	0.00570
	(0.0306)	(0.0128)
Year indicator: 2013	−0.0562*	−0.0253*
	(0.0330)	(0.0138)
Year indicator: 2014	−0.0331	−0.0216
	(0.0370)	(0.0170)
Constant	9.651***	9.198***
	(0.682)	(0.154)
Observations	2,820,723	2,820,723

NOTE: SEs are in parentheses. *** $p < 0.01$. ** $p < 0.05$. * $p < 0.1$. Shading highlights that, although Figure 4.4 in Chapter Four shows that the 1943 birth cohort has a substantial increase in federal pay between ages 35 and 40, the result is not statistically significantly different from 0.

Using the birth cohort models, we find similar results to the cross-sectional models that we estimated in the last section, including that wages rise sharply at younger ages; however, several important differences exist. First, we observe in Figure 4.4 in Chapter Four that the 1943 birth cohort has a substantial increase in federal pay between ages 35 and 40, but this result is not statistically significantly different from 0 (as observed in Table A.2, shaded). However, pay remains persistently higher for the 1943 birth cohort after 40 (as well as the 1948 birth cohort), suggesting that something might be characteristically different about these federal workers that the model does not capture. For example, it could be that these workers were able to retain greater pay rates during the period of high inflation in the early 1980s than new entrants into the federal civil service could. Education for the 1963 birth cohort is associated with similar patterns to what was observed in the cross-section—namely, that professional degrees yield the highest return, followed by doctorates and master's degrees. Interestingly, although the return to doctorates and master's degrees do not vary significantly by birth cohort, the return to professional degrees in the private sector exhibits a significant decline over younger birth cohorts. There is no clear downward trend in the federal sector. This might suggest that an oversupply of professional degrees has diminished wages for these jobs in the private sector. There are also interesting differences in the association between pay and veteran status. Although, for the 1963 birth cohort, veteran status is associated with lower pay in both the federal and private sectors, the net effect is zero or positive for veterans born before 1951 in the private sector. This time frame corresponds to the end of the draft and suggests that pre- and postdraft veteran status might capture employment characteristics that we do not otherwise capture in our model.

Entry Cohort–Specific Dynamic Retention Models with Fixed Personal Discount Factors

To explore how mean taste varies across entry cohort holding constant the personal discount factor, we reestimated the models in Tables 5.1 and 5.2 in Chapter Five with a fixed value of the discount factor. The value we chose was the estimated personal discount factor in the DRM that combines data for entry cohorts 1993 to 2000. As we show later in this appendix, the estimated personal discount factor in the combined model is 0.93. Tables B.1 through B.3 report the parameter estimates of these models with fixed personal discount factors.

To highlight how the estimated parameters of distribution of tastes differ when the personal discount factor is fixed, we reproduce Figure 5.2 from Chapter Five but insert the results of the models for which the personal discount factor is fixed. Figure B.1 shows the new results. Comparing Figures B.1 and 5.2, it is clear that, in the earlier entry cohorts of 1988 and 1989, both mean tastes and the SDs of taste show less variability across entry cohort when the personal discount factor is fixed. For example, estimated mean taste for the 1988 cohort is $10,280, falling to –$8,875 for the 1992 cohort when the personal discount factor is estimated and free to vary. In contrast, when the personal discount factor is fixed, estimated mean taste for the 1988 entry cohort is only $701 and not statistically significantly different from 0 and falls to –$4,679 only for the 1992 entry cohort.

Table B.1
Parameter Estimates and Standard Errors for Models with Fixed Personal Discount Rates, by Entry Cohort, 1988 Through 1994

Parameter	1988		1989		1990		1991		1992		1993		1994	
	Estimate	SE	Estimate	SE	Estimate	SE	Estimate	SE	Estimate	SE	Estimate	SE	Estimate	SE
Estimated location, λ	60.519	2.54	60.202	1.96	56.127	2.80	48.978	2.48	44.557	2.55	48.043	3.98	41.94	3.13
Mean of nonveterans' taste distribution, μ	0.701	0.51	1.220	0.39	0.119	0.54	−2.857	0.40	−4.679	0.41	−6.662	0.60	−7.384	0.55
Mean of veterans' taste distribution, μ	0.389	0.88	0.303	0.67	0.303	0.78	0.118	1.04	1.923	0.85	5.150	1.15	4.422	0.84
SD of nonveterans' taste distribution, σ	16.316	1.03	13.547	0.71	12.781	1.04	11.58	0.99	12.61	1.14	16.687	1.92	16.205	1.70
SD of veterans' taste distribution, σ	−9.542	2.21	−5.813	1.56	−8.140	2.56	0.065	1.80	−2.030	1.64	−3.595	2.19	−4.223	1.64
Accounting for the censoring until 1995, θ	−2.662	0.08	−3.319	0.10	−3.208	0.13	−2.418	0.09	−3.701	0.27	−2.838	0.18	−3.541	0.28
Shift in the probability of staying in the first year due to early attrition, δ	−2.646	0.09	−2.696	0.07	−3.085	0.13	−3.321	0.19	−3.313	0.23	−3.432	0.35	−4.041	0.60
Personal discount factor, β	2.588	0	2.588	0	2.588	0	2.588	0	2.588	0	2.588	0	2.588	0
Log likelihood	−18,165		−31,233		−14,828		−13,618		−10,574		−6,885		−7,410	
Sample size	7,363		12,758		6,167		5,638		4,495		2,976		3,296	

Table B.2
Parameter Estimates and Standard Errors for Models with Fixed Personal Discount Rates, by Entry Cohort, 1995 Through 2000

Parameter	1995		1996		1997		1998		1999		2000	
	Estimate	SE	Estimate	SE	Estimate	SE	Estimate	SE	Estimate	SE	Estimate	SE
Estimated location, λ	44.096	3.64	45.201	3.93	42.364	4.39	45.710	5.67	65.790	13.79	48.704	6.61
Mean of nonveterans' taste distribution, μ	−6.358	0.69	−3.647	0.70	−3.442	0.80	−3.060	0.94	5.209	3.12	1.066	1.52
Mean of veterans' taste distribution, μ	3.544	0.87	1.126	0.88	2.245	1.05	1.976	1.01	0.503	1.54	3.095	1.14
SD of nonveterans' taste distribution, σ	17.990	2.06	18.542	2.18	17.993	2.70	17.324	2.99	26.069	6.68	17.828	3.1712
SD of veterans' taste distribution, σ	−5.432	1.88	−3.070	1.95	−2.685	2.34	−6.360	2.65	−11.354	5.42	−6.498	3.04
Accounting for the censoring until 1995, θ	−3.999	0.52		0		0		0		0		0
Shift in the probability of staying in the first year due to early attrition, δ	−4.009	0.63	−4.410	0.71	−3.448	0.41	−3.125	0.28	−2.718	0.18	−3.022	0.18
Personal discount factor, β	2.588	0	2.588	0	2.588	0	2.588	0	2.588	0	2.588	0
Log likelihood	−7,577		−6,683		−5,492		−4,545		−4,825		−6,379	
Sample size	3,513		3,249		2,808		2,327		2,729		3,810	

Table B.3
Transformed Parameter Estimates in Models with Fixed Personal Discount Factors, by Entry Cohort

Parameter	1988	1989	1990	1991	1992	1993	1994	1995	1996	1997	1998	1999	2000
Estimated location, λ	60.52	60.20	56.13	48.98	44.56	48.04	41.94	44.10	45.20	42.36	45.71	65.80	48.70
Mean of nonveterans' taste distribution, μ	0.70	1.22	0.12	−2.86	−4.68	−6.66	−7.38	−6.36	−3.65	−3.44	−3.06	5.21	1.07
Mean of veterans' taste distribution, μ	0.39	0.30	0.30	0.12	1.92	5.15	4.42	3.54	1.13	2.24	1.98	0.50	3.10
SD of nonveterans' taste distribution, σ	16.32	13.55	12.78	11.58	12.61	16.69	16.20	17.99	18.54	17.99	17.32	26.07	17.83
SD of veterans' taste distribution, σ	−9.54	−5.81	−8.14	0.06	−2.03	−3.59	−4.22	−5.43	−3.07	−2.69	−6.36	−11.35	−6.50
Accounting for the censoring until 1995, θ	0.07	0.03	0.04	0.08	0.02	0.06	0.03	0.02	0.00	0.00	0.00	0.00	0.00
Shift in the probability of staying in the first year due to early attrition, δ	0.07	0.06	0.04	0.03	0.04	0.03	0.02	0.02	0.01	0.03	0.04	0.06	0.05
Personal discount factor, β	0.93	0.93	0.93	0.93	0.93	0.93	0.93	0.93	0.93	0.93	0.93	0.93	0.93
Percentage change in force size as a result of a 1% across-the-board decrease in real pay	−3.2	−3.8	−3.9	−5.1	−4.5	−3.5	−4.1	−3.8	−3.5	−3.2	−3.7	−2.2	−2.9

Figure B.1
Estimated Parameters of the Taste Distribution When the Personal Discount Factor Is Fixed, by Entry Cohort

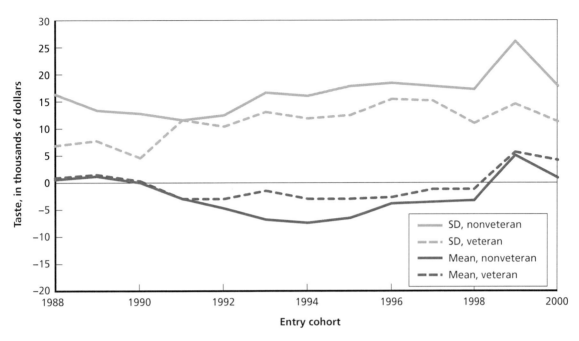

References

Asch, Beth J., James Hosek, and Michael G. Mattock, *A Policy Analysis of Reserve Retirement Reform*, Santa Monica, Calif.: RAND Corporation, MG-378-OSD, 2013. As of August 28, 2016:
http://www.rand.org/pubs/monographs/MG378.html

―――, *Toward Meaningful Military Compensation Reform: Research in Support of DoD's Review*, Santa Monica, Calif.: RAND Corporation, RR-501-OSD, 2014. As of August 28, 2016:
http://www.rand.org/pubs/research_reports/RR501.html

Asch, Beth J., Michael G. Mattock, and James Hosek, *The Federal Civil Service Workforce: Assessing the Effects on Retention of Pay Freezes, Unpaid Furloughs, and Other Federal-Employee Compensation Changes in the Department of Defense*, Santa Monica, Calif.: RAND Corporation, RR-514-OSD, 2014a. As of August 28, 2016:
http://www.rand.org/pubs/research_reports/RR514.html

―――, *How Do Federal and Civilian Pay Freezes and Retirement Plan Changes Affect Employee Retention in the Department of Defense?* Santa Monica, Calif.: RAND Corporation, RR-678-OSD, 2014b. As of August 28, 2016:
http://www.rand.org/pubs/research_reports/RR678.html

―――, *Reforming Military Retirement: Analysis in Support of the Military Compensation and Retirement Modernization Commission*, Santa Monica, Calif.: RAND Corporation, RR-1022-MCRMC, 2015. As of August 28, 2016:
http://www.rand.org/pubs/research_reports/RR1022.html

Chu, David S. C., and John P. White, "Ensuring Quality People in Defense," in Ashton B. Carter and John Patrick White, eds., *Keeping the Edge: Managing Defense for the Future*, Cambridge, Mass.: MIT Press, 2000, pp. 203–234.

Congressional Budget Office, *Characteristics and Pay of Federal Civilian Employees*, Pub. 2839, March 1, 2007. As of February 15, 2016:
http://www.cbo.gov/publication/18433

Deaton, Angus, "Panel Data from Time Series of Cross-Sections," *Journal of Econometrics*, Vol. 30, No. 1–2, October–November 1985, pp. 109–126.

Falk, Justin, *Comparing Wages in the Federal Government and the Private Sector*, Congressional Budget Office, Working Paper 2012-3, January 30, 2012. As of August 28, 2016:
https://www.cbo.gov/publication/42922

Federal Retirement Thrift Investment Board, *Thrift Savings Plan: Analysis of Participant Behavior and Demographics for 2009–2013*, undated. As of September 1, 2016:
https://www.frtib.gov/ReadingRoom/SurveysPart/Behavior-Demographics-2013.pdf

Gates, Susan M., Edward G. Keating, Adria D. Jewell, Lindsay Daugherty, Bryan Tysinger, Albert A. Robbert, and Ralph Masi, *The Defense Acquisition Workforce: An Analysis of Personnel Trends Relevant to Policy, 1993–2006*, Santa Monica, Calif.: RAND Corporation, TR-572-OSD, 2008. As of September 11, 2016:
http://www.rand.org/pubs/technical_reports/TR572.html

Gotz, Glenn A., and John McCall, *A Dynamic Retention Model for Air Force Officers: Theory and Estimates*, Santa Monica, Calif.: RAND Corporation, R-3028-AF, 1984. As of August 28, 2016:
http://www.rand.org/pubs/reports/R3028.html

Isaacs, Katelin P., *Federal Employees' Retirement System: Benefits and Financing*, Washington, D.C.: Congressional Research Service, January 30, 2014. As of September 3, 2016:
http://digitalcommons.ilr.cornell.edu/key_workplace/1220/

Lunney, Kellie, "Republican Budget Asks Federal Employees to Contribute More to Pensions," *Government Executive*, April 1, 2014. As of August 28, 2016:
http://www.govexec.com/oversight/2014/04/budget-story/81676/

Mattock, Michael G., and Jeremy Arkes, *The Dynamic Retention Model for Air Force Officers: New Estimates and Policy Simulations of the Aviator Continuation Pay Program*, Santa Monica, Calif.: RAND Corporation, TR-470-AF, 2007. As of August 28, 2016:
http://www.rand.org/pubs/technical_reports/TR470.html

Mattock, Michael G., Beth J. Asch, James Hosek, Christopher Whaley, and Christina Panis, *Toward Improved Management of Officer Retention: A New Capability for Assessing Policy Options*, Santa Monica, Calif.: RAND Corporation, RR-764-OSD, 2014. As of August 28, 2016:
http://www.rand.org/pubs/research_reports/RR764.html

Office of Personnel Management, "Executive Branch Civilian Employment Since 1940," undated (a). As of April 25, 2016:
https://www.opm.gov/policy-data-oversight/data-analysis-documentation/federal-employment-reports/historical-tables/executive-branch-civilian-employment-since-1940/

———, *Veterans Employment Initiative: Vet Guide*, undated (b). As of February 21, 2016:
https://www.opm.gov/policy-data-oversight/veterans-employment-initiative/vet-guide/

OPM—*See* Office of Personnel Management.

Pew Research Center, "Comparing Millennials to Other Generations: Male Labor Force Status When They Were Ages 18–33," March 19, 2015. As of April 24, 2016:
http://www.pewsocialtrends.org/2015/03/19/comparing-millennials-to-other-generations/#!6

PricewaterhouseCoopers, *Millennials at Work: Reshaping the Workplace*, 2011. As of April 24, 2016:
https://www.pwc.com/gx/en/managing-tomorrows-people/future-of-work/assets/reshaping-the-workplace.pdf

Public Law 112-96, Middle Class Tax Relief and Job Creation Act of 2012, February 22, 2012. As of August 31, 2016:
https://www.gpo.gov/fdsys/pkg/PLAW-112publ96/content-detail.html

Stafford, Darlene E., and Henry S. Griffis, *A Review of Millennial Generation Characteristics and Military Workforce Implications*, Alexandria, Va.: CNA, CRM D0018211.A1/Final, May 2008. As of September 11, 2016:
https://www.cna.org/CNA_files/PDF/D0018211.A1.pdf

Thrift Savings Plan, "C Fund: Common Stock Index Investment Fund," undated (a). As of April 25, 2016:
https://www.tsp.gov/InvestmentFunds/FundOptions/fundPerformance_C_Perf.html

———, "G Fund: Government Securities Investment Fund," undated (b). As of April 25, 2016:
https://www.tsp.gov/InvestmentFunds/FundOptions/fundPerformance_G.html

Tobin, James, "Estimation of Relationships for Limited Dependent Variables," *Econometrica*, Vol. 26, No. 1, January 1958, pp. 24–36.

TSP—*See* Thrift Savings Plan.

U.S. Census Bureau, "Current Population Survey," undated. As of September 2, 2016:
http://www.census.gov/programs-surveys/cps.html

———, "Annual Social and Economic Supplement (ASEC) of the Current Population Survey (CPS)," last revised December 9, 2015. As of September 2, 2016:
https://www.census.gov/did/www/saipe/data/model/info/cpsasec.html

van der Klaauw, Wilbert, and Kenneth I. Wolpin, "Social Security and the Retirement and Savings Behavior of Low-Income Households," *Journal of Econometrics*, Vol. 145, No. 1–2, July 2008, pp. 21–42.

Wiatrowski, William J., "The Last Private Industry Pension Plans: A Visual Essay," *Monthly Labor Review*, December 2012, pp. 3–18. As of August 28, 2016:
http://www.bls.gov/opub/mlr/2012/12/art1full.pdf

Zawodny, Ken, "FERS-Revised Annuity Employee, FERS-RAE," *Retirement Information Center*, May 13, 2013. As of September 3, 2016:
https://www.opm.gov/blogs/Retire/2013/5/13/FERS-Revised-Annuity-Employee-FERS-RAE/